AF342516

Current Ornithology

Current Ornithology

Series editor: **Charles F. Thompson**
Illinois State University, Normal, Il, USA

For other titles published in this series, go to
www.springer.com/series/5816

Charles F. Thompson

Editor

Current Ornithology

Volume 17

 Springer

Editor
Charles F. Thompson
Illinois State University
Normal, Il, USA
wrens@ilstu.edu

ISSN 0742_390X
ISBN 978-1-4419-6420-5   e-ISBN 978-1-4419-6421-2
DOI 10.1007/978-1-4419-6421-2
Springer New York Dordrecht Heidelberg London

Printed on acid-free paper

Springer is part of Springer Science+Business Media (www.springer.com)

Contents

Contributors

Robert H. Day
ABR, Inc., Environmental Research and Services, P.O. Box 80410,
Fairbanks, AK 99708, USA

Mark A. Fraker
TerraMar Environmental Research Ltd., 8617 Lochside Drive, Sidney,
BC, Canada V8L 1M8

Sven Jakobsson
Department of Zoology, Stockholm University, S-106 91 Stockholm, Sweden

Cecilia Kullberg
Department of Zoology, Stockholm University, S-106 91 Stockholm, Sweden

Johan Lind
Centre for the Study of Cultural Evolution, Department of Zoology,
Stockholm University, 106 91 Stockholm, Sweden
johan.lind@zoologi.su.se

Stephen M. Murphy
ABR, Inc., Environmental Research and Services, P.O. Box 80410,
Fairbanks, AK 99708, USA

Christopher M. Perrins
Department of Zoology, Edward Grey Institute of Field Ornithology, University
of Oxford, South Parks Road, Oxford OX1 3PS, UK
chris.perrins@zoology.oxford.ac.uk

S. James Reynolds
Centre for Ornithology, School of Biosciences, College of Life & Environmental
Sciences, University of Birmingham, Edgbaston, Birmingham, B15 2TT, UK
J.Reynolds.2@bham.ac.uk

David L. Swanson
Department of Biology, University of South Dakota, 414 East Clark Street,
Vermillion, SD 57069, USA
david.swanson@usd.edu

John A. Wiens
PRBO Conservation Science, 3820 Cypress Drive #11, Petaluma, CA 94954, USA
jwiens@prbo.org

Chapter 1
Impaired Predator Evasion in the Life History of Birds: Behavioral and Physiological Adaptations to Reduced Flight Ability

Johan Lind, Sven Jakobsson, and Cecilia Kullberg

1.1 Introduction

Because the failure to escape a predator causes death to a prey animal, and thus excludes opportunities to reproduce in the future, predation is a major selective force in nature (e.g., Lima and Dill 1990; Dawkins and Krebs 1979). To evade attacks from an array of different predators successfully is thus of key importance for all potential prey organisms. In most animals there are periods in their life when they are more susceptible to predation than at other times. For example, the reproductive period might be associated with enhanced predation risk (see Magnhagen 1991 for a review). In birds, the most common way of escaping from predators is to use the ability to fly, and when birds are attacked by predators, take-off ability and maneuverability in flight are crucial for survival (e.g., Rudebeck 1950). Changes in body mass or wing area will change wing load (body mass/wing area, see Pennycuick 1989) and thus potentially will affect flight ability. Such changes in wing load may lead to variation in predation risk during a bird's life. Periods when birds may have impaired evasive abilities because of changes in wing load include during migration (increased mass from fat storage), reproduction (increased mass from egg load), and molt (reduced wing area from feather loss and growth).

An impaired ability to evade predator attacks may be subdivided into two parts (Witter and Cuthill 1993): (1) reduced take-off ability and (2) impaired performance in flight once fully airborne. When discussing flight in general, we will use the term "flight ability," which incorporates both of these aspects of predator evasion (1 and 2). To emphasize the importance of rapid take-off, in terms of both speed and angle of ascent, it should be noted that many predators rely on surprise attack, where the success rate is much reduced if the prey gets fully airborne. Furthermore, take-off ability may affect an individual's survival, because many

J. Lind (✉)
Centre for the Study of Cultural Evolution, Department of Zoology,
Stockholm University, 106 91 Stockholm, Sweden
e-mail: johan.lind@zoologi.su.se

C.F. Thompson (ed.), *Current Ornithology Volume 17*, Current Ornithology,
DOI 10.1007/978-1-4419-6421-2_1, © Springer Science+Business Media, LLC 2010

avian species evade surprise attacks by short, quick flights to protective cover (Cresswell 1996; Lindström 1989; Newton 1986; Kenward 1978; Rudebeck 1950). Recently, Bednekoff (1996) suggested that predation risk per attack increases exponentially with the time required to reach protective cover, so high flight speed should be important when escaping from predator attacks. But another theoretical study suggests that under some circumstances flight ability may be of minor importance for birds subjected to predators' surprise attacks (Lind 2004). Another important aspect of predator evasion is maneuverability (e.g., Howland 1974); some birds of prey are not solely dependent on surprise attacks, but may pursue the prey for considerable time and distance (e.g., Rudebeck 1950). As these aspects of flight are important in terms of predator avoidance, they most likely influence individual survival and fitness; hence, it is reasonable to believe that they have been optimized through natural selection (e.g., Lima 1993).

Empirical studies so far suggest that birds face an increased predation risk through reduced ability to escape when wing load increases. Table 1.1 summarizes the empirical studies that we refer to in this review. The effects of increases in body mass on escape flight ability are summarized in Fig. 1.1. The scope of this review is to discuss research that has investigated how various factors influence bird-flight ability, especially in the context of evading predators. Because increased predation risk from impaired predator evasion may lead to different adaptations, we will focus on the physiological and behavioral adaptations that reduce the risk of being killed.

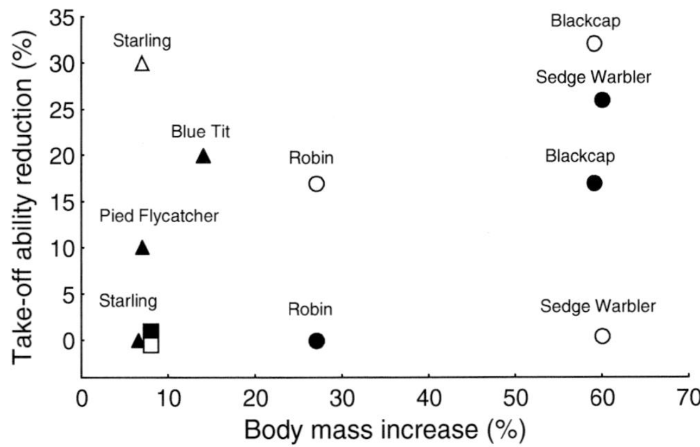

Fig. 1.1 Impaired take-off ability (%) in relation to natural body mass increase (%) due to reproduction, diurnal body mass increase and migratory fuel load. *Filled symbols* indicate effect on take-off speed and *open symbols* indicate effects on take-off angle. *Filled triangle/open triangle* – reproduction. *Filled square/open square* – diurnal body mass increase. *Open circle/ filled circle* – migratory fuel load. Symbols for diurnal body mass increase represents six sedentary where no effect could be found. Data from studies in Table 1.1

Table 1.1 Empirical studies of flight ability in birds referred to in this study

Variation studied	Variation achieved	Species	Measured effects on flight ability	Experimental setup	References
Daily mass	Natural	Great Tit	No (EF) No (RF)	1 1	Kullberg et al. (1998)
Daily mass	Natural	Willow Tit	No (EF) No (RF)	1 1	Kullberg (1998)
Daily mass	Natural	Zebra Finch	Speed (RF)	3	Metcalfe and Ure (1995)
Daily mass	Natural	Zebra Finch	No (EF)	3	Veasey et al. (1998)
Daily mass	Natural	Yellowhammer	No (EF)	1	van der Veen and Lindström (2000)
		Greenfinch	No (EF)	1	
Daily mass	Manipulated	Starling	Maneuverability and angle of ascent (EF)	2 2	Witter et al. 1994
Migratory fuel	Natural	Blackcap	Speed and angle of ascent (EF)	1 1	Kullberg et al. (1996)
Migratory fuel	Natural	Robin	Angle of ascent (EF)	1	Lind et al. (1999)
Migratory fuel	Natural	Sedge Warbler	Speed (EF)	1	Kullberg et al. (2000)
Egg load	Natural	Starling	Angle of ascent (EF)	2	Lee et al. (1996)
Egg load	Natural	Blue Tit	Speed (EF)	3	Kullberg et al. (2002a)
Pectoral muscle and egg load	Natural	Zebra Finch	Speed (EF)	3	Veasey et al. (2000a, b)
Pectoral muscle and mass during incubation	Natural	Pied Flycatcher	Speed (EF)	3	Kullberg et al. (2002b)
Pectoral muscle and egg load	Natural	Zebra Finch	Speed and angle of ascent (EF)	1	Kullberg et al. (2005)
Molt	Natural Manipulated	Starling	Angle of ascent (EF) Speed and maneuverability (EF)	2 2	Swaddle and Witter (1997)
Molt	Natural Manipulated	Tree Sparrow	Non (EF) Speed (EF)	1 1	Lind 2001
Pectoral muscle, body mass during molt	Natural and manipulated	Tree sparrow	Speed	3	Lind and Jakobsson (2001)

(continued)

Table 1.1 (continued)

Variation studied	Variation achieved	Species	Measured effects on flight ability	Experimental setup	References
Asymmetry	Natural	Starling	Speed (EF)	2	Swaddle et al. (1996)
	Manipulated		Maneuverability (EF)	2	
Asymmetry	Natural	Starling	Angle of ascent (EF)	2	Swaddle and Witter (1997)

EF escape flight (the birds are in various ways startled to take-off); *RF* routine flight (the birds take-off undisturbed). Studies of escape flight has been categorized according to experimental setup: *1* take-off when attacked by a model raptor; *2* take-off when released directly from the hand of the experimenter accompanied by a startle sound; *3* take-off in a narrow cage, forcing the bird to fly vertically by releasing it from the hand or from a box

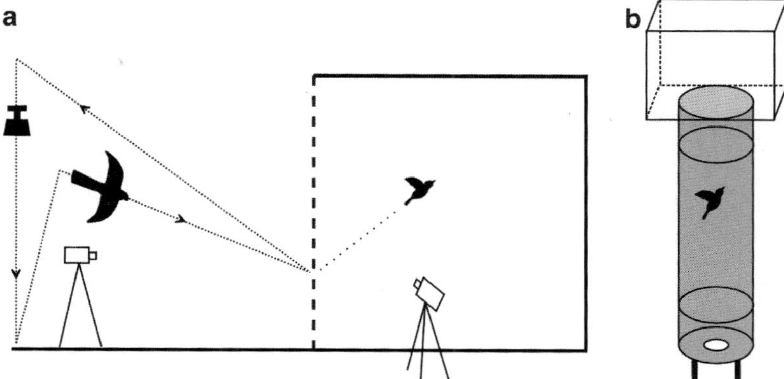

Fig. 1.2 Examples of methods used to measure escape flights. (**a**) Experimental setup where a predator model emerges from a blind simulating a predator attack. (**b**) Vertical take-off ability has been measured in a high narrow cage. The birds are released at the bottom and trapped in the collecting cage at the top

1.2 Measuring Escape Flight Ability

Empirical studies measuring variation in escape flight ability in birds have used different setups to record the initial take-off within the first 1.5 m of the flight. The take-offs are video recorded and then analyzed with respect to speed and, in some cases, angle of ascent. We have broadly defined the studies cited in this review into three different categories (Table 1.1):

1. Take-off when attacked by a model raptor (Fig. 1.2a)
2. Take-off when released directly from the hand of the experimenter accompanied by a startle sound
3. Take-off in a narrow cage, forcing the bird to fly vertically by releasing it from the hand or from a box (Fig. 1.2b)

In addition, maneuverability has been assessed in some studies by flying birds through an obstacle course with wooden poles hanging from the roof of a long narrow aviary. Maneuverability has been measured by counting the number of contacts with the poles and time required to fly across the course (see Witter et al. 1994).

1.3 Body Mass Variation

The relationship between increased body mass and flight ability was the first issue studied in the framework of impaired predator evasion. The suggestive and indirect evidence for a cost of increasing mass by increasing fat reserves (in terms of increased risk of predation) came from the fact that birds store energy well below physiological and environmental maxima (e.g., Macleod et al. 2007; Gosler et al. 1995;

Witter and Cuthill 1993; Rogers 1987; Blem 1976). As pointed out by Witter and Cuthill (1993), predation risk, metabolism, risk of injuries, foraging efficiency, reproduction, and pathology may be affected by high body mass, thereby reducing the individual's fitness. Theoretically, increased body mass could affect predation risk in mainly two ways: indirectly by mass-dependent metabolism and directly by mass-dependent predation risk (Houston and McNamara 1993). A heavy bird may have higher energy demands for maintenance of body mass that may force it to spend more time foraging, and, thus, suffer from increased exposure to predators (e.g., Witter and Cuthill 1993; Lima 1986). Furthermore, a heavy bird may be less adept at escaping from a predator once it is actually attacked (e.g., Hedenström 1992; Blem 1975).

The most widely recognized reason for birds varying their body mass is through variation in fat reserves. Birds use fat as their main source of energy, both to ensure overnight survival and as fuel during migration (e.g., Blem 1976). Furthermore, during reproduction female birds increase in body mass because of the presence of an egg load and the growth of the gonads (Moreno 1989). Several theoretical studies have assumed that increased body mass should affect flight ability detrimentally by increasing wing load (e.g., Pennycuick 1989; Lima 1986; Blem 1976). Hedenström (1992) substantiated this assumption by using bird-flight theory (Pennycuick 1975), and showed theoretically that many aspects of bird flight, such as minimum turning radius, acceleration, maximum flight speed, and climb rate, should be negatively affected by an increase in body mass.

1.3.1 *Diurnal Variation in Body Mass*

The regulation of energy reserves in small birds during winter has received much attention in recent years. Fat, particularly triglycerids, makes up most of birds' energy reserves and is efficiently used as energy storage because oxidation of fat yields more energy per unit mass than either carbohydrate or protein (Schmidt-Nielsen 1990). However, energy reserves do contain more than just fat (Dolnik and Bluymental 1967). For example, proteins have been shown to make up a significant part of birds' energy reserves during long-distance migration (e.g., Bauchinger and Biebach 2001; Lindström and Piersma 1993). Therefore, the term fuel load instead of fat load is preferably used when discussing energy reserves (see also a review by Piersma and Lindström 1997). These energy reserves may be regulated by a trade-off between the benefits (reduced starvation risk) and the costs of increased risk of predation (see, e.g., Lind and Cresswell 2005; McNamara and Houston 1990; Lima 1986). According to several theoretical studies, wintering birds should adjust daily energy reserves to food availability, metabolic requirements, and predation risk (Brodin 2000; Houston et al. 1993, 1997; Clark and Ekman 1995; Bednekoff and Houston 1994a, b; Grubb and Pravosudov 1994; McNamara et al. 1994; McNamara and Houston 1990; Lima 1986). If food is scarce or unpredictable and if metabolic requirements are high because of, for example, low ambient temperatures, birds should increase energy reserves to reduce the risk of starvation. Empirical studies of the regulation of energy reserves in wintering birds support these theoretical predictions (e.g., Macleod et al. 2007; Gosler 1996; Lilliendahl et al. 1996; Bednekoff and Krebs 1995; Witter et al. 1995; Ekman and Hake 1990; Rogers 1987; however, see Dall and Witter 1998). Furthermore, in accordance with the assumption that enlarged energy reserves increase the risk of predation, some studies indicate that wintering birds adjust body mass to the perceived risk of predation, which suggests a trade-off between the risk of starvation and predation. In England, a population of Great Tits (*Parus major*) became heavier over a period of years when Sparrowhawks (*Accipiter nisus*) were absent because of pesticide poisoning, and declined in mass when the hawks were reestablished (Gosler et al. 1995; but see Cresswell et al. 2009). Similarly, in a laboratory study of wintering Greenfinches (*Carduelis chloris*) a stuffed predator model that moved in the bird's environment resulted in lower body masses (Lilliendahl 1997; see also van der Veen 1999). But birds have also been found to increase their body mass in response to increased predation risk (Lilliendahl 1998; Pravosudov and Grubb 1998). A large-scale study of 30 bird species and over 300,000 banded individuals showed that the way that birds respond with mass change to predation risk seems to be determined by habitat quality and the opportunities for safe foraging conditions (Macleod et al. 2007; see also Cresswell et al. 2009).

In the first empirical study actually investigating flight performance in relation to natural increase in body mass, Metcalfe and Ure (1995) showed impaired flight in Zebra Fiches (*Taeniopygia guttata*) attributable to a 7% increase in body mass during the day. Zebra Finches had a 30% lower take-off speed at dusk than at dawn. The authors concluded from their results that small birds are inherently more vulnerable to predation at dusk than at dawn. However, the result was mainly based on spontaneous take-offs without predator stimuli and only a few measurements were made on alarmed take-offs. Spontaneous take-offs are likely to be affected by a heavy body mass because birds might fly slower when heavier to reduce energy costs (Pennycuick 1989). On the other hand, when attacked by a predator birds should use all available energy to minimize the risk of being caught (cf. Hedenström and Alerstam 1995). So far empirical studies on escape flight ability in birds have been unsuccessful in demonstrating any change in escape flight as a result of natural increases in daily body mass ranging from 6 to approximately 8% (e.g., studies on Yellowhammers (*Emberiza citrinella*) and Greenfinches: van der Veen and Lindström 2000; Willow Tits (*Parus montanus*): Kullberg 1998: Great Tits – Kullberg et al. 1998; Zebra Finches: Veasey et al. 1998). In contrast to these findings, Witter et al. (1994) found an effect on maneuverability and angle of ascent in European Starlings (*Sturnus vulgaris*) when adding artificial weights in the range of 7–14% of the birds' body mass. This is in the normal range of variation in diurnal mass; however, the use of artificial weights may make interpretations ambiguous because the center of gravity may be altered and drag may be increased in a way that induces additional and unwanted effects on take-off and on maneuverability (Metcalfe and Ure 1995). In the same experiment (Witter et al. 1994) natural body mass was manipulated by food deprivation before trials, resulting in an experimental group on average 5.3 g (±0.64 g) lighter than the control birds. The food-deprived, and therefore lighter, birds leaped more steeply and ascended at steeper angles than controls during take-off, suggesting increased take-off ability caused by the reduction in body mass brought about by the food deprivation. Moreover, the food-deprived birds were more maneuverable than the controls when flying through an obstacle course.

Thus, the available evidence suggests that the modest energy reserves accumulated during a day in, for example, wintering birds does not seem to increase predation risk as a consequence of reduced flight ability. This does not, however, exclude the possible importance of indirect mass-dependent predation risk in birds attributable to daily mass increase (e.g., Adriaensen et al. 1998; Gosler et al. 1995). Theoretical work predicts that an increase in body mass should incur an increased indirect mass-dependent predation risk caused by increased exposure to predators during the time spent foraging (cf. Lind 2004; Lima 1985, 1986).

1.3.2 *Migratory Fuel Load*

In contrast to the modest fuel loads stored during winter nights in sedentary birds, migratory birds store large quantities of fuel, sometimes exceeding 100% of lean body mass (e.g., Fry et al. 1970). Although these extreme fuel loads occur mostly when birds are about to cross wide geographic barriers, fuel loads of 20–30% are commonly found in birds during migration (cf. Alerstam and Lindström 1990). Blackcaps (*Sylvia atricapilla*) that breed in northern Europe are tropical migrants that store large fuel loads during migration. Kullberg et al. (1996) showed that take-off ability of Blackcaps was impaired by migratory fuel load when they were exposed to a simulated attack by a predator model. Both take-off speed and take-off angle were affected, and it was estimated that a Blackcap increasing its fuel load from 0 to 60% would take off at a 32% lower angle and a 17% slower speed. In addition, a study of Robins (*Erithacus rubecula*) showed that these medium-distance migrants also suffer from reduced take-off ability (Lind et al. 1999). Although take-off speed was not affected at all, a fuel load of 27% (the fuel load of the heaviest bird in that study) led to a 17% shallower angle at take-off compared with a bird carrying no fuel load. A study of take-off ability in Sedge Warblers (*Acrocephalus schoenobaenus*), just prior to their Trans-Saharan crossing, indicates that Sedge Warblers as opposed to Robins seem to favor maintaining their angle of ascent at the expense of speed. Sedge Warblers carrying a 60% fuel load flew 26% slower than birds carrying no fuel load, but their angle of ascent was unaffected (Kullberg et al. 2000). In the take-off, birds probably face a trade-off between speed and angle of ascent because a low angle permits the highest acceleration. These results suggest that take-off decisions differ between species (cf. Kullberg et al. 1996; Witter et al. 1994).

In conclusion, migratory fuel load may increase the risk of predation in birds through impaired predator evasion and not only because of increased exposure when foraging to accumulate these large energy reserves. But studies suggest that heavy birds can compensate behaviorally which means that heavy birds do not have to suffer from augmented mortality due to their large fuel loads (Cimprich et al. 2005; Lind and Cresswell 2006; Lind 2004).

1.3.3   Reproduction

During the reproductive season female birds increase in body mass by 7–30% through gonad growth and egg production (see Moreno 1989 for references), which in turn may adversely affect their ability to escape from predators. Furthermore, in addition to the effect of increased mass on wing loading, the eggs may change the center of gravity of the bird toward the poster, which may affect both aerial maneuverability and flight speed (see Srygley and Dudley 1993). An increased body mass from carrying an egg may, therefore, affect flight ability more than that caused by an increased fuel load, which can be deposited subcutaneously around the body in a way that has less of an effect on the center of gravity than does an egg. An early study that focused on the reproductive behavior of males in Bank Swallows (*Riparia riparia*) found that males can detect heavy females because of the females' impaired flight at the time of egg laying and use this as a cue for extra-pair copulation attempts (Beecher and Beecher 1979). This suggests that the flight ability of female Bank Swallows is impaired during the egg-laying period, when they may increase in body mass by as much as 20%. Jones (1986) substantiated this by experimentally increasing body mass by intraperitoneal injection of saline water (2 g of water was injected, corresponding to an increase in body mass of approximately 15%). The "weighted" birds took longer to reach ascending flight from release compared with control birds. The experimentally "weighted" birds did not, however, differ in flight ability from the naturally heavy (gravid) females, thus showing that the impaired flight ability was caused by the increase in body mass. In accordance with this, Lee et al. (1996) found large effects on flight ability during simulated predator escape in gravid European Starlings. Female European Starlings were, on average, about 7% heavier, and their take-off angle was 30% lower during escape when carrying eggs than it was before egg production. Furthermore, the first study of flight ability in wild birds during reproduction indicates that the loading effect during egg laying impairs flight ability (Kullberg et al. 2002a). Individual Blue Tits (*Cyanistes caeruleus*) were captured during the egg-laying and early chick-rearing stages of the nesting cycle. At each trapping occasion body mass, flight muscle size, and flight ability were measured (for nonterminal measurement of flight muscle size, see Selman and Houston 1996). Females were 14% heavier during egg laying than during the early chick-rearing stage while male body mass did not change. There was no change in flight muscle size in either males or females, leading to a lower flight muscle ratio (flight muscle size/body mass) for females during egg laying. As a consequence, female Blue Tits flew 20% slower than males during egg laying. However, during the early chick-rearing stage, when female birds had reduced their body mass and had the same flight muscle ratio as males, there was no difference in flight ability between the sexes.

There seems to be an additional factor beyond that imposed by the direct effect of egg load that can affect flight ability during reproduction in female birds. In some bird species, the mass of eggs produced by a female over a few days may exceed the female's body mass. Recent studies have shown that in a wide range of bird species females reduce flight muscle size during the laying period (Houston et al. 1995b; Jones 1990). It has been suggested and also shown experimentally in captive Zebra Finches (Houston et al. 1995c), that this decline represents a direct contribution of muscle proteins to egg production (Houston et al. 1995a, c; Jones and Ward 1976; Kendall et al. 1973). Because the flight muscle ratio positively correlates with flight ability (Marden 1987), one might thus expect a reduced take-off ability caused by a decrease in pectoral muscle during reproduction in female birds. However, muscle sarcoplasm in the pectoral muscle may act as a protein reserve that can be depleted without impairing the contractile function of the muscle (Houston et al. 1995b; Kendall et al. 1973). Recently, Veasey et al. (2000a) experimentally manipulated the physiological cost of egg production in female Zebra Finches by varying both the number of eggs a female laid and the quality of her pre-laying diet. The removal of the first eggs in a clutch will induce the laying of a larger clutch than normal and thus encourage the female to invest more energy in egg laying. In another study, Veasey et al. (2000b) manipulated clutch size of individual female Zebra Finches. The same individual thus experienced both small and large clutch sizes in a series of successive breeding attempts. The results from both these studies (Veasey et al. 2000a, b) showed that experimentally enlarged costs during egg laying in female Zebra Finches resulted in a large reduction in flight muscle. This led to reduced flight ability after completion of the clutch compared with flight ability before start of egg laying. However, a low physiological cost of reproduction caused only a small reduction in muscle size and females actually flew faster after completion of the clutch than before start of egg laying. These results are in accordance with the suggested occurrence of protein storage in the flight muscle. Small reductions in flight muscle size in birds laying a small clutch might represent a reduction of these reserve proteins and may not involve a decrease in muscle function but instead lead to improved flight ability through a reduced wing load. On the other hand, females facing a large physiological cost of reproduction as a result of experimentally increased clutch size may be forced to use the contractile proteins in the flight muscle, which in turn will reduce their flight ability.

Not only investment in eggs and egg load might affect flight ability in birds during reproduction. During incubation both male and female birds might weigh more because of high energy demands. This has recently been shown to affect vertical take-off ability adversely in wild breeding female Pied Flycatchers (*Ficedula hypoleuca*). A 7% reduction in body mass after the chicks had hatched was associated with a 10% increase in vertical take-off speed compared with that during incubation (Kullberg et al. 2002b.)

1.4   Molt

The period of molt, the shedding and replacement of old feathers, may affect bird flight because the loss of flight feathers will reduce wing area and increase wing loading (Ginn and Melville 1983). Many birds, for example divers (Gaviidae), grebes (Podicepedidae), ducks and geese (Anatidae), become flightless during molt (Ginn and Melville 1983) when they shed simultaneously all of their flight feathers. This is normally performed at sites that provide protection from predators (Ginn and Melville 1983) and where these species are normally not dependent on flight for predator evasion. Although flightlessness may also occur in fast-molting passerines (Haukioja 1971), most passerines remain capable of flight during molt and continue to rely on flight to evade predators. Yet, the possibility that molt constrains flight in birds has received little attention, especially in a framework of increased risk of predation. Hedenström and Sunada (1999) showed theoretically that the greatest effect of molt on flight ability should occur during midmolt, and that large molt gaps should be more detrimental than smaller gaps; thus, both the position and the size of the molt gaps influence flight ability (Fig. 1.3). An interesting finding, however, was that these theoretically estimated effects of molt gaps were rather small.

The first study to investigate experimentally the effect of molt on flight ability found that molt impairs flight ability in European Starlings (Swaddle and Witter 1997). European Starlings in escape flights during natural molt took off at lower

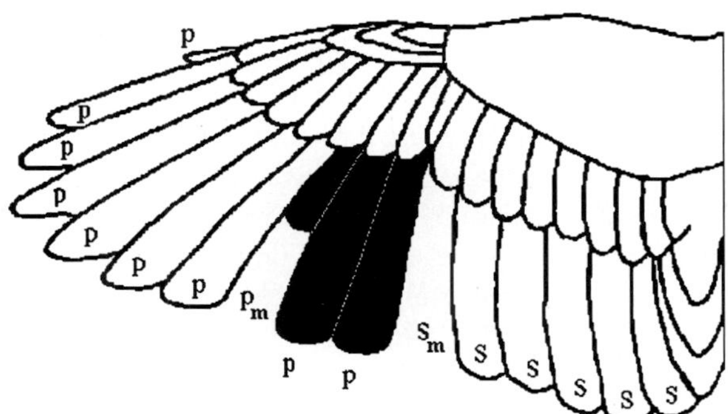

Fig. 1.3   A passerine wing in molt (*p* primary; *s* secondary). Primaries and secondaries are collectively called remiges. Dark feathers are new feathers (fully-grown or in growth). The areas where feathers have been shed and not yet been completely replaced are referred to as molt gaps. When birds molt, the shedding of remiges result in molt gaps. Molt gaps are indicated by p_m and s_m

angles than nonmolting birds. As expected from the work of Hedenström and Sunada (1999), the effect was largest during midmolt, when the angle of ascent was approximately 33% lower than at the onset of molt. However, the authors did not find any effect on take-off speed in naturally molting birds. In the same study, Swaddle and Witter (1997) experimentally simulated molt in European Starlings, which resulted in reduced take-off speed, lower level flapping flight speed, and reduced maneuverability during escape flights; however, molt simulation did not affect take-off angle.

The importance of the position of the molt gap was demonstrated in an experiment by Chai (1997). When investigating the energetics of hovering, he found that Ruby-throated Hummingbirds (*Archilochus colubris*) were more affected when molting the primaries than when molting the secondaries. This is not in agreement with the theoretical work of Hedenström and Sunada (1999), who predicted that the effect should be largest when the molt gap occurs well inside the wing. However, this may be explained by the fact that molt of primaries incurs a reduction in wing span, which is critical for lift production during hovering flight (see also Chai and Dudley 1999). Chai et al. (1999) also found that Ruby-throated Hummingbirds are constrained during molt in their maximal horizontal flight speed, most likely caused by the reduced maximal power output from which the molting birds suffered.

In contrast to those studies showing effects on flight ability attributable to the presence of molt gaps, a study of escape flight in molting Eurasian Tree Sparrows (*Passer montanus*) suggests that for this and other slowly molting species the effect of the small molt gaps might be less on flight ability and thus predation risk (Lind 2001). No effect was found on take-off speed or angle of ascent in naturally molting Eurasian Tree Sparrows that had a maximum of 2.5 missing remiges per wing. However, when larger molt gaps of up to ten missing remiges were simulated on the Eurasian Tree Sparrows a large detrimental effect on take-off speed (18%) occurred. This result is contrary to the theoretical study by Hedenström and Sunada (1999), where the effect of molt on flight was predicted to be surprisingly small.

Wings may not only be affected by gaps during molt; uneven feather regrowth during molt or abrasion could also cause wings to become asymmetric (Ginn and Melville 1983; Francis and Wood 1989). Such morphological asymmetries may have implications for the performance of birds in flight (see Norberg 1990 for a review). The first experimental study of this issue showed that asymmetry might affect different aspects of flight in the European Starling (Swaddle et al. 1996). Naturally occurring asymmetry on primaries caused by damage was correlated with lower angle of ascent in escaping European Starlings, but no correlation with take-off speed could be found. To eliminate any confounding effects (e.g., poor flyers could have more damaged feathers), Swaddle et al. (1996) also manipulated primary length and asymmetry. Increased asymmetry did not affect take-off ability but did impair maneuverability. Reducing primary length, on the other hand, reduced take-off ability in terms of take-off speed but not maneuverability. Because the manipulations resulted in lengths and asymmetries outside the natural range of fully-grown feathers in European Starlings, and were more similar to patterns occurring during molt, the authors suggest these findings may have implications for molting strategies.

1.5 Adaptations to Compensate for Impaired Predator Evasion

As reviewed so far, birds face several periods in their lives when ability to evade predators might be impaired by reduced flight ability. If predation risk, as well as the individually perceived predation risk, increases through impaired ability to evade predators, we would expect selection for physiological and behavioral adaptations that reduce the risk of being killed during these periods of increased predation risk.

1.5.1 Adaptive Reversible Changes in Physiology Caused by Reduced Flight Ability

When birds face an increased risk of predation because of higher wing load, which will impair predator evasion, several morphological changes in the birds' bodies can occur to improve predator evasion. The ability to change the size of various organs adaptively has until recently been underestimated in birds, but recent research has revealed patterns of organ plasticity during different life-history stages. This plasticity in the regulation of the size and mass of body organs cannot, of course, be regarded only as an adaptation to reduce predation risk by maintaining escape ability. Organ plasticity may also act to maintain optimal flight ability during migration (Lindström et al. 2000; Biebach 1998) and during times of high work loads (Bautista et al. 1998; Swaddle and Biewener 2000; Moreno 1989; Norberg 1981). Muscle hypertrophy may also result from protein storage in anticipation of protein loss during long migratory flights (Biebach 1998; Jenni and Jenni-Eiermann 1998), or for use on arrival at the breeding grounds (Fransson and Jakobsson 1998; Davidson and Evans 1988; McLandress and Raveling 1981).

During migration birds increase in body mass as they increase their fuel load, which results in increased wing loading. Flight muscle hypertrophy, an increase in muscle mass, may be one way to compensate for the detrimental effects of increased wing load. This hypertrophy seems to be a general phenomenon in migrating birds (Biebach 1998; Piersma 1998; Gaunt et al. 1990; Marsh 1984; Fry et al. 1972). For example, the flight muscles of Gray Catbirds (*Dumetella carolinensis*) may increase as much as 35% during fall premigratory fattening (Marsh 1984). This plasticity is not restricted to the flight muscles. A study of Great Knots (*Calidris tenuirostris*) suggests that during long-lasting flights (in this case 5,400 km) no organs apart from lungs and brains remain homeostatic (Battley et al. 2000). Moreover, by reducing leg mass, visceral mass (e.g., Jehl 1997), and organs such as gonads, liver, and intestines, birds might reduce body mass and thus wing load in preparation for migration (for a review, see Piersma and Lindström 1997). For example, reduction of the digestive tract occurs in passerine long-distant migrants prior to migration, and it has been suggested that these changes in organ size are very rapid and may occur in the course of a few days (Hume and Biebach 1996; Klaassen and Biebach 1994).

Birds need energy to produce eggs. Yet, most passerine bird species are probably able to accumulate energy and nutrients on a more or less daily basis for egg production (Meijer and Drent 1999; Woodburn and Perrins 1997; Perrins 1996). But, birds relying on diets low in protein, such as the Zebra Finch, may use protein from their pectoral muscle for egg production and thus, under some conditions (manipulated enlarged clutch size), experience a reduced flight ability after the completion of the clutch (Veasey et al. 2000a, b). In Zebra Finches, the occurrence of a protein reserve in the pectoral muscle may be an adaptation to avoid contractile muscle depletion during egg production. In a recent study of take-offs using a model predator, nonbreeding female Zebra Finches flew just as slowly as females with a natural egg load. Nonbreeding females retained this slow take-off speed throughout the study while egg-laying females increased their flight speed as their body mass was reduced after they completed their clutch (Fig. 1.4). If female Zebra Finches store protein when the opportunity presents itself to enable a rapid onset of breeding, they pay a cost in terms of reduced flight ability prior to egg laying (Kullberg et al. 2005).

Molting birds may also compensate physiologically for their impaired flight ability. Body mass decreases during molt in both Ruby-throated Hummingbirds and European Starlings (Chai 1997; Swaddle and Witter 1997, respectively), resulting in a possibly compensatory reduced wing load. Generally, molt is associated with a low level of fat (Dolnik and Gavrilov 1979 and references therein) and low body mass compared with other times during the year, as shown in the studies on European Starlings (Swaddle and Witter 1997) and Ruby-throated Hummingbirds (Chai 1997). However, as molt commences, birds face an increase in water content in the body as a result of, for example, degradation, redistribution, and resynthesis of proteins and an increased blood volume perfusing active feather pulps. This increase in water results in a slight increase in body mass during the period of molt; nevertheless, body mass remains low during molt (see, e.g., Flinks and Kolb 1997;

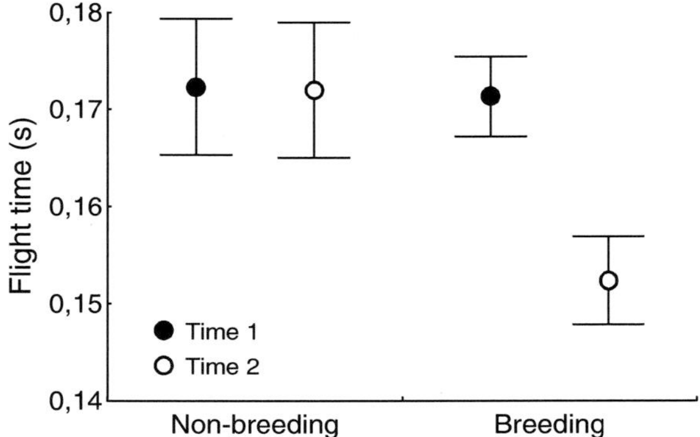

Fig. 1.4 Time to fly 40 cm during escape take-off (mean ± SE) from a simulated predator attack for nonbreeding ($n = 12$) and breeding ($n = 12$) zebra finches. Birds were subjected to the experiment once when the breeding females laid their first egg (time 1) and once more after the last egg of the clutch was laid (time 2) (data from Kullberg et al. 2005)

Lindström et al. 1994; Dolnik and Gavrilov 1979). Interestingly, molt is also associated with an increase in daily variation of nonfat components of the body (Dolnik and Gavrilov 1979). The low fat content that occurs during molt may be a strategic body mass reduction; however, reductions are most likely concurrently constrained by, for example, increased water content of their bodies (Lind et al. 2004).

Another way of increasing flight ability when facing a reduced wing area during molt would be to increase flight muscle. Dissections of Yellow-vented Bulbuls (*Pycnonotus goiavier;* Ward 1969) indicate reductions in lean dry weight of flight muscles during molt. However, the only study so far on intra-individual change in flight muscle during molt shows that Eurasian Tree Sparrows increase the ratio between pectoral muscle size and body mass as the wing area is reduced during molt. Subsequently, as the molt is completed and the wing area increases, this ratio is reduced (Lind and Jakobsson 2001). This possibly explains the lack of empirical evidence for reduced flight ability in molting Eurasian Tree Sparrows (Lind 2001). Furthermore, a novel pattern of strategic organ flexibility was found when Eurasian Tree Sparrows were subjected to a simulated molt pattern outside the molting period. In response to a reduction in wing area, Eurasian Tree Sparrows reduced their body mass concurrently with an increase in pectoral muscle size, which shows that the regulation of body mass and flight muscle is effectively unlinked (Lind and Jakobsson 2001). This is in contrast to earlier studies on migratory birds that found positive correlations between muscle size and body mass (e.g., Bautista et al. 1998; Lindström et al. 2000; Swaddle and Biewener 2000; Biebach 1998; Piersma 1998). Pectoral muscle size has also been shown to track changes in body mass during incubation in female Pied Flycatchers (Kullberg et al. 2002b). Female pectoral muscle size was reduced concurrently with the reduction in wing loading after nestlings hatched. That pectoral muscle size actually can covary with predation risk per se has been shown experimentally in Ruddy Turnstones (*Arenaria interpres*), which increased their pectoral muscle size after predation risk was manipulated through presentations of a flying raptor model (van den Hout et al. 2006).

1.5.2 Adaptive Behavioral Changes Caused by Reduced Flight Ability

When birds suffer from reduced flight ability and thus may be subject to a higher predation risk, it is likely that they alter their behavior adaptively to reduce this risk (for review, see Lind and Cresswell 2005). Newton (1966) suggested that molting birds do adjust their behavior in order to avoid avian predators. Captive Bullfinches (*Pyrrhula pyrrhula*) in molt were active only while feeding and spent the rest of each day sitting in shaded positions, in contrast to their lively activity pattern when not molting (Newton 1966). Passerines in heavy molt may become more secretive, and hence less likely to catch the eye of predators (Haukioja 1971). For example, Swaddle and Witter (1997) showed that manipulated molt causes European

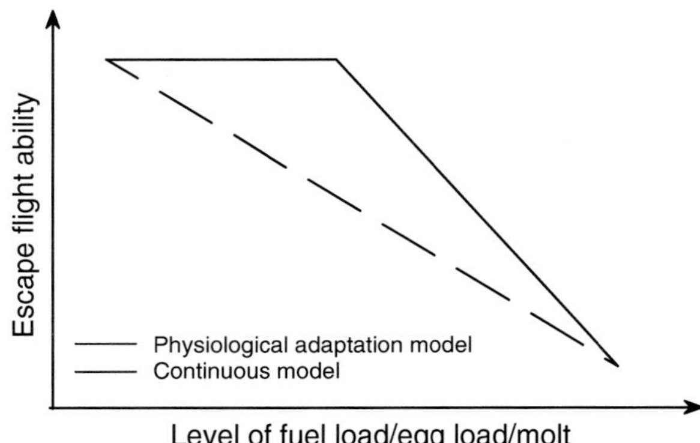

Fig. 1.5 Relationship between escape flight and variation in body composition. A heuristic graph showing the likely effect physiological adaptations to increased wing load has on the effect on escape flight. It is compared with the conventional models that describe the relationship between increase in wing load and flight ability as being continuous. The rationale for suggesting a threshold model lies in the empirical findings showing that escape flight may be unaffected by small body mass increase and slow molt

Starlings to become less active and seek relatively more protective areas. Similar behavior occurred when the flight ability of European Starlings was detrimentally affected by attached artificial weights (Witter et al. 1994). Furthermore, during reproduction, when flight ability in female birds is impaired by egg load and reduced flight muscle mass, female Zebra Finches alter their behavior to reduce predation risk. During the egg-laying period females reduced feeding time, avoided foraging at high risk patches, and tended to associate with larger flocks when feeding (Veasey 1999).

Boreal birds in winter are constrained by the trade-off between the risk of starvation and predation (e.g., McNamara and Houston 1990). During long, cold winter nights they rely on fat accumulated during the day for survival. Daily energy reserves in wintering parids range from 7 to 15% of the birds' lean body mass (Haftorn 1989; Lehikoinen 1987). As we reported above in this review, no empirical study has so far been able to measure any effect of a natural increase of daily body mass in the range of 6–8% of lean body mass on take-off ability in escaping birds. Behaviors such as hoarding and hypothermia may be adaptations in wintering birds, not only to minimize risk of starvation, but also to avoid large fuel loads that may detrimentally affect flight. Food caches can be considered as external fuel storage that has no effect on flight ability. Hence, they represent fuel storage with no cost associated with direct mass-dependent risk of predation (McNamara et al. 1990). Furthermore, by lowering body temperature 3–10°C during the night, wintering parids can reduce energy consumption and thereby reduce the energy reserves needed for the night (Reinertsen and Haftorn 1984; Haftorn 1972). However, deep hypothermia may increase predation risk during the night by reducing the ability of the bird to react to nocturnal predators (Bednekoff et al. 1994; Reinertsen and Haftorn 1984).

During migration, birds are forced to put on fat as fuel for migratory flight (Fig. 1.6). Normally, migratory birds put on fuel loads of 20–30% of lean body mass and make several stopovers to forage and replenish energy reserves during their journey to the wintering grounds. Decisions about the amount of fuel load and timing of departure from a stopover site probably depend on factors like time, food

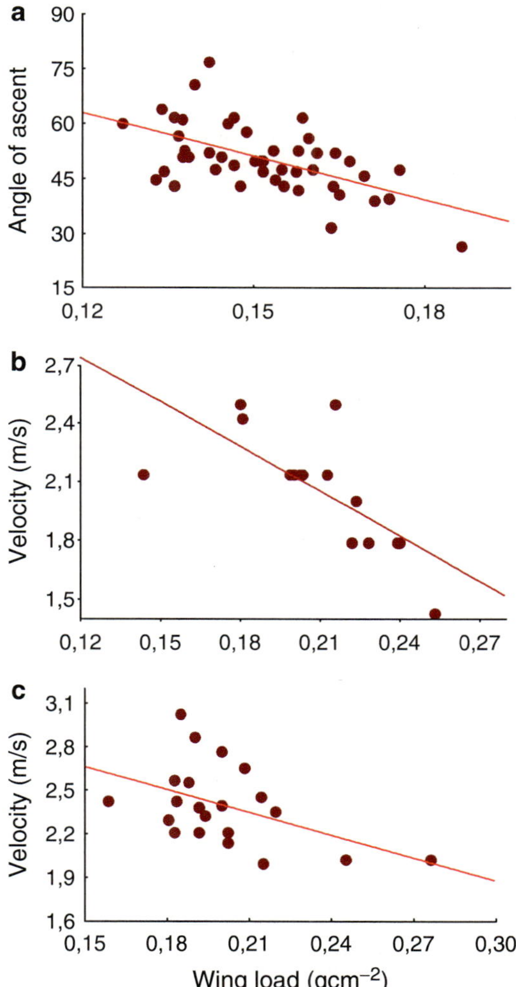

Fig. 1.6 Empirical studies of how migratory fuel load affect escape flight. (a) Robins, a medium-distance migrant, took off less steeply with increasing fuel loads ($r^2 = 0.32$, $p = 0.00004$, $n = 46$). Fuel load ranged between 0 and 27%; (b) Sedge warblers' take-off speed is negatively affected by migratory fuel load ($r^2 = 0.45$, $p = 0.004$, $n = 15$). Fuel load ranged between 0 and 67%; (c) take-off speed in Blackcaps is negatively affected by migratory fuel load ($r^2 = 0.22$, $p = 0.03$, $n = 21$). Fuel load ranged between 0 and 59%. (a) Is published with permission from Springer (Fig. 2 in Lind et al. 1999) and (b) with permission from AOU (Fig. 1d in Kullberg et al. 2000)

availability, and predation risk (Alerstam and Lindström 1990; Fransson 1997). In a laboratory experiment, Fransson and Weber (1997) showed that Blackcaps adjusted their rate of fuel deposition and stopover time to the perceived risk of predation at the stopover site. Blackcaps with a high risk of predation increased food intake rate and showed a pattern of night activity indicating that they wanted to leave the simulated stopover site earlier and with a lower fuel load than a control group. This finding is consistent with experimental studies on Yellow-rumped Warblers (*Dendroica coronata*) that showed that the warblers resumed feeding faster when in migratory disposition than did control birds after exposure to a predator model (Moore 1994). This suggests that birds are more prone to predation risk when building up migratory fuel loads. In addition, Moore and Simms (1986) showed that Yellow-rumped Warblers foraged in a risk-prone manner during the actual fuelling period, whereas they foraged in a risk-averse manner when they had taken on their fuel load and had begun to show nocturnal restlessness. That birds become predation-risk prone during migratory fuelling has also been shown in Ruddy Turnstones (*Arenaria interpres*), which decrease their level of antipredatory behavior when fuelling (Metcalfe and Furness 1984).

1.6 Discussion and Prospects

Results from experiments on flight ability in birds have until now produced just as many questions as answers about impaired predator evasion. However, the number of studies performed and the even smaller number of species studied make it clear that to get a better understanding of the field more research is needed. Additional studies of other species with different life histories are needed to assess the general applicability of the results obtained to date. In the sections that follow, we consider the issues that we have identified as being of special interest for future studies, and we also make an attempt to elucidate some of the problems associated with previous studies.

1.6.1 Methodology

A critical assumption in all studies on escape flight ability is that birds in the experimental setup perform maximally. However, the variation in the methods used to startle the birds in the studies we have cited may affect the response of the bird. For example, a bird released from the hand and exposed to a startling noise may react differently from a bird that is actually attacked and pursued by a model raptor. Furthermore, some of the methods used force the birds to ascend vertically while others give the bird the opportunity to choose a take-off strategy trading off speed and angle of ascent. Thus, there is a need to evaluate different methods, for example, by using the same set of birds in different experimental setups.

There is growing evidence for fast functional physiological responses to short-term changes of wing load in birds. In one experiment, European Starlings that had the opportunity to get used to artificially added weights for 3 days before a take-off experiment performed better than birds that were subject to newly added weights (Witter et al. 1994). There was also a tendency for a compensatory reduction in natural body mass during the course of these 3 days. In another study, take-off speed of European Starlings was restored to pretreatment levels 2 weeks after a simulated molt was induced (Swaddle and Witter 1997). These improvements in flight ability could, of course, be the result of behavioral adaptations to fly with a heavy wing load, or to use wings with reduced area. However, in response to an experimentally reduced wing load, which simulated molt, Eurasian Tree Sparrows were able to change body mass within 5 days and muscle size within 7 days (Lind and Jakobsson 2001). Thus, when subjecting birds to experimental manipulation to resemble natural variation in wing load, one has to bear in mind that these fast responses may occur. Furthermore, besides the change in organ size during different life-history stages that we mentioned earlier in this review, there are reasons to believe that variation in other traits, such as water content and hormonal levels, might influence the magnitude of the effect on flight ability through, for example, experimentally increased wing load.

Nonterminal methods (e.g., ultrasound, body tomographic imaging techniques, external measurements) that give the opportunity to study changes in different body parts within individuals could be used to yield valuable information. The recently developed technique of measuring flight muscle size in birds by using dental alginate is a simple method that can be used under field conditions and yields highly repeatable results (Selman and Houston 1996). Interpretation of changes in the size of flight muscles, however, requires that we need to learn more about muscle composition and the possible relevance of protein storage in the flight muscle (Houston et al. 1995b). If there is such protein storage, reduction in muscle size could occur without a concurrent reduction in contractile function and power output.

1.6.2 *Migration*

Long-distance migrants face many extreme situations during migration. Considering the number of birds involved in Transcontinental migration, surprisingly little has been done to study the effect of predation risk during migration. Until now, impaired predation evasion has been empirically studied in only a few species, all which have shown negative effects of migratory fuel load on take-off ability (Burns and Ydenberg 2002; Kullberg et al. 1996, 2000; Lind et al. 1999). The studies on fuelling and foraging in birds during migration (Fransson and Weber 1997; Moore 1994; Moore and Simms 1986; Metcalfe and Furness 1984) suggest that birds change behaviorally, becoming more risk prone when fuelling at stopover sites. Considering that birds during migration spend a majority of the time at stopover sites (approximately 70–80% of the time during migration for *Sylvia* spp. warblers;

Fransson 1995), the fuelling periods are of critical importance. The increased risk-taking, together with impaired escape ability because of fuel loads, probably leads to an enhanced predation risk. However, a recent theoretical study shows that the risk incurred by increased wing loading, such as migratory fuel load, can be compensated for behaviorally (Lind 2004). Possible behavioral adaptations that reduce predation risk when carrying a high fuel load are, of course, not easy to document in wild, small birds. However, with the recent improvements of small radio transmitters (see, for example, Naef-Daenzer 1993), it might be possible to overcome these difficulties. Therefore, we think that newly developed techniques can be very useful for investigating the relationship between predation risk and behavioral adaptations during migratory stopover. It could also be possible to study behavioral patterns in captive birds during different stages of fuel loading (see, for example, Fransson and Weber 1997; Moore 1994; Moore and Simms 1986). There is much more to learn about strategic changes in body composition during migratory stopover and flights (for a review, see Piersma and Lindström 1997).

1.6.3 Reproduction

Variation in flight ability during reproduction in female birds has only been studied in three species, and more studies are obviously needed to reveal the generality of patterns observed. There are reasons to expect species-specific differences in the effect of reproduction on flight ability because of the great variation in reproductive strategies. Among the factors that may affect these differences are whether or not females deposit body reserves before reproduction or use daily food intake for the production of eggs (capital or income breeders; Stearns 1992). Furthermore, the occurrence of supplementary feeding from the male and variation in the cost of incubation and brooding and provisioning of young may influence the magnitude of any effect on flight ability, and hence the possibility for increased predation risk caused by an impaired ability to evade predators.

Only one study has investigated change in flight ability in males during reproduction (Kullberg et al. 2002a). Although female Blue tits suffered from impaired flight ability attributable to increased wing loading during egg laying, males flew just as fast during the egg-laying period as during the early chick-rearing period. However, in some species, males invest in incubation and build up energy reserves (Moreno 1989). How these reserves affect flight ability remains to be investigated. In species where males' testes sizes are very large, males might face a trade-off between testes size and flight ability during the mating season. For example, male testes size in the Alpine Accentor (*Prunella collaris*) represents about 8% of body mass (Nakamura 1990), which is similar to the gain in mass that a gravid females may experience (Lee et al. 1996).

As suggested by Veasey et al. (2000a, b), maternal predation risk caused by reduced flight ability during reproduction might be relevant in a trade-off between reproductive effort and survival. Since an experimental increase in clutch size can reduce flight muscle condition in female birds (Veasey et al. 2000a, b; Monaghan et al. 1998), predation risk may be an important factor affecting clutch size in birds, and could be one explanation why birds often lay lower clutch sizes than optimality theories have predicted (see Monaghan and Nager 1997, for a review). However, the effect of an enlarged clutch size on flight ability in birds has only been studied in captive Zebra Finches. Female Zebra Finches rely on stored protein and other body reserves for production of eggs (Houston et al. 1995a), which probably is an adaptation to their low protein diet (Houston et al. 1995a). However, most small passerines are able to accumulate energy and nutrients on a more or less daily basis for egg production (Meijer and Drent 1999; Woodburn and Perrins 1997; Perrins 1996). For example, in Blue Tits it has been shown that supplementary, i.e., "court-ship," feeding of the female by the male is equivalent to her daily energy require-ments for egg production (Krebs 1970). Furthermore, in a study of change in body composition in female Blue Tits during reproduction, there was no decline in flight muscle mass during the egg-laying period (Woodburn and Perrins 1997), indicating that Blue Tits might not face the same trade-off between clutch size and flight ability as Zebra Finches.

Concerning possible adaptations to costs of large clutch size (in terms of reduced flight ability), brood parasites, such as cuckoos and cowbirds, would be an interest-ing group to study. In this group of birds there are species that produce more eggs than any other wild species (Davies 2000). For example, female Brown-headed Cowbirds *(Molothrus ater)* lay about 40 eggs during an 8-week breeding period and are thus about five times as productive as bird species investing in parental care (Scott and Ankney 1983). Furthermore, female cuckoos and cowbirds may have to fly long distances during their egg-laying period to find host nests. In the Brown-headed Cowbird a female's breeding area could be as large as 60 times the size of that of a similar-sized species in which the female cares for its own young (Rothstein et al. 1984).

1.6.4 Molt and Wing Morphology

There is a large variation in patterns of wing molt in birds. The duration of molt depends on the life history of the species (Jenni and Winkler 1994). Long-distance migrants in temporal stress, brought about by the need to commence migration, generally molt very fast, usually in approximately 30–50 days (Jenni and Winkler 1994). Such species, therefore, suffer a greater reduction in wing area (i.e., have larger molt gaps in their wings) during molt than species that molt more slowly. For example, Thrush Nightingales (*Luscinia luscinia*) complete molt in approximately 35 days; Ginn and Melville 1983). In contrast, sedentary species generally have a slower molt than migratory species resulting in much smaller molt gaps. Resident passerine species complete molt in 60–85 days, whereas partial migrants take 50–85 days to complete molt (Jenni and Winkler 1994). Thus, migratory species with a postnuptial molt (i.e., a molt between the period of reproduction and fall migration) probably face a trade-off between fast molt and increased predation risk caused by reduced flight ability during molt (cf. Haukioja 1971). Slowly molting sedentary species, on the other hand, are able to minimize the negative effects on aerodynamic qualities of the wings and thereby maintain their flight ability largely unaffected, as indicated by a study on flight ability in Eurasian Tree Sparrows (Lind 2001). Predation risk might thus be an important factor affecting molt strategies, and we encourage further investigations of molt strategies in relation to predation risk in birds (see Slagsvold and Dale 1996).

The newly found adaptive change in body mass and muscle size during molt in Eurasian Tree Sparrows (Lind and Jakobsson 2001) likely explains why no effect of natural molt on escape flight ability has been found in this species (Lind 2001). However, how general these adaptive changes are, and if these changes in muscle volume also occur in migratory birds during molt, remains to be investigated (see discussion in Lind et al. 2004). For instance, given that compensation is costly, it is possible that skulking species with no need to fly during molt do not compensate physiologically, whereas species living in the open and relying on their flight ability to evade predators and to catch prey (e.g., flying insects) do compensate physiologically for their wing load increase (cf. Lind 2004).

In many passerine species, juvenile birds have shorter and more slotted wings than adults (Alatalo et al. 1984), which has been suggested to result from nutritional constraints (e.g., Slagsvold 1982). However, Alatalo et al. (1984) suggested that this difference is an adaptation to increase survival of young birds. Short, slotted wings lead to slower speed but to increased maneuverability in flight compared with longer wings (Pennycuick 1975; Savile 1957). Alatalo et al. (1984) suggest that young birds might trade-off maneuverability against speed to increase such skills as feeding efficiency and predator avoidance, while older birds that can fly faster might compensate for their reduced maneuverability with experience. This merits further investigation.

1.6.5 *Implications for Theoretical Studies*

There has been a great interest in describing bird flight theoretically (e.g., Hedenström and Sunada 1999; Norberg 1990, 1995; Hedenström 1992; Rayner 1988; Pennycuick 1969, 1975). For example, the cost of flight increases with increasing wing load; however, there is no flight theory developed for take-off, nor how take-off ability is affected by an increase in wing load. The same is true for alarmed take-offs, where birds perform near their maximum capacity. Moreover, theories applicable to predator evasion are scarce. Howland (1974) modeled the situation in which a predator chases a prey, labeled the turning gambit. The model investigates what affects that the maneuverability and speed of the predator relative to the prey has on the outcome of the chase. The model is applicable to, for example, a falcon hunting birds in mid-air, but not to the major hunting strategy in avian predators, namely hunting by surprise attacks. The turning gambit was expanded by

Hedenström and Rosén (2001), who examined idealized attack-escape situations (escape by climbing, horizontal speeding, turning and diving). Intraspecific variation (representing intra-individual variation) in body mass, muscle size and wing area was excluded from their analysis. Nevertheless, a similar approach may prove fruitful in examining the effect of variation in body composition on escape performance in birds. This could generate predictions regarding behavioral shifts, including, for example, when birds should change antipredator behavior (both escape behavior and predation risk averse/prone foraging) caused by the inevitable body mass increase during a stopover period.

Bednekoff (1996) did model the effect of escape speed (explicitly, time to reach cover) on survival for attacked prey birds; unfortunately, his model has not yet been tested. However, one might need to reevaluate the generality of this model by incorporating scan rates other than those used, because the effect shown in the model is strongly dependent on high scan rates. Hence, the model is most applicable to, and should be tested on, singletons with high scanning frequencies. A similar theoretical study based on other scan rates also showed that predation risk is determined by an interaction between fuel load and distance to cover (Lind 2004). Furthermore, we think that it could be fruitful to describe formally optimal escape behavior in a surprise attack situation, which may aid empiricists in interpreting data and designing experiments.

A general pattern in the empirical studies described in this review is that moderate increase in wing load does not affect flight ability detrimentally (Lind 2001; van der Veen and Lindström 2000; Veasey et al. 1998; Kullberg 1998; Kullberg et al. 1998; but see Witter et al. 1994). In contrast, during reproduction, moderate increases in body mass have been shown to affect flight ability negatively (Lee et al. 1996; Kullberg et al. 2002a, b). Furthermore, large fuel loads in migratory birds and reduced wing area during molt have also been shown to affect escape ability (Lind et al. 1999; Kullberg et al. 1998; Swaddle and Witter 1997). Taken together, these results suggest that birds are able to compensate for moderate increase in wing load by body alterations (for example by increasing flight muscle volume and reduction in body mass; Lind and Jakobsson 2001). This indicates that the relationship between escape flight ability and wing load may not be continuous (e.g., Hedenström 1992; Norberg 1990) but might be better described by a threshold model (Kullberg et al. 1998; Lind 2001) (Fig. 1.5).

1.6.6 Applied Aspects of Predator Evasion

Whether or not birds are affected by body mass increase does also have some practical applications. As a rule of thumb, radio transmitters used on birds should weigh less than 5% of body mass (e.g., Naef-Daenzer 1993). The studies of escape take-off discussed in preceding sections strongly indicate that within this range flight ability is not negatively affected in small passerines. However, to the best of our knowledge no experimental studies have been performed to evaluate flight ability in relation to added weights in form of, for example, radio transmitters. Moreover, it is likely that a bird's center of gravity (compare Sects. 3.3 and 6.3 about reproduction and studies where experimental weights have been used, e.g., Witter et al. 1994) is affected when a radio transmitter is attached, which may lead to effects not comparable to fuel loads in the same mass range. Nevertheless, from the studies presented in this review, we conclude that the traditionally chosen level (<5% of the bird's body mass) is reasonable, especially in situations where transmitters are glued on the birds. When attaching a transmitter using a harness, impairment of flight may appear for reasons other than the effect of an increase in body mass (Gessaman et al. 1991; however, see Rappole and Tipton (1991) for harnesses that work well on small passerines).

1.6.7 Concluding Remarks

Even though our knowledge of how various life-history stages affect escape flight ability in birds is increasing, fitness consequences of impaired predator evasion are still difficult to quantify. To determine fitness consequences of increased wing loading, or any other behavior affecting predation risk, one needs to consider that individuals have the possibility to compensate for apparently risky behaviors by decreasing their risk, or increasing their reproductive output, by means of alternative compensatory behaviors (Lind and Cresswell 2005). Ideally, birds varying in flight ability should be observed when attacked by predators in the wild; however, predatory events are hard to observe and quantify. Nevertheless, field studies have been successful in showing how important the behavioral responses of the prey are during predator attacks (e.g., Cresswell 1993) and also in determining how relative predation risk varies between species (e.g., Solonen 1997; Götmark and Post 1996). Yet, studies focusing on the predator, in the framework we have presented here are still scarce. We suggest that focusing on predator behavior may be necessary to advance our knowledge of impaired predator evasion in the life history of birds.

Acknowledgment We thank Franz Bairlein, Innes Cuthill, Thord Fransson, Neil Metcalfe, Charles F. Thompson, and an anonymous referee for valuable comments on the manuscript. We are especially grateful for stimulating discussions with Innes Cuthill, Thord Fransson, David Houston, Will Cresswell, Åke Lindström, and Neil Metcalfe. C.K. was funded by a fellowship from the Swedish National Research Council.

References

Adriaensen F, Dhondt AA, VanDongen S, Lens L, Matthysen E (1998) Stabilizing selection on Blue Tit fledgling mass in the presence of Sparrowhawks. Proc R Soc Lond B 265: 1011–1016

Alatalo RV, Gustafsson L, Lundberg A (1984) Why do young passerine birds have shorter wings than older birds? Ibis 126:410–415

Alerstam T, Lindström Å (1990) Optimal bird migration: the relative importance of time, energy, and safety. In: Gwinner E (ed) Bird migration. Springer, Berlin, pp 47–67

Battley PF, Piersma T, Dietz MW, Tang SX, Dekinga A, Hulsman K (2000) Empirical evidence for differential organ reductions during trans-oceanic bird flight. Proc R Soc Lond B 267:191–195

Bauchinger U, Biebach H (2001) Differential catabolism of muscle protein in Garden Warblers (*Sylvia borin*): flight and leg muscles act as a protein source during long-distance migration. J Comp Physiol B 171:293–301

Bautista LM, Tinbergen J, Wiersma P, Kacelnik A (1998) Optimal foraging and beyond: how starlings cope with changes in food availability. Am Nat 152:543–561

Bednekoff PA (1996) Translating mass dependent flight performance into predation risk: an extension of Metcalfe and Ure. Proc R Soc Lond B 263:887–889

Bednekoff PA, Houston AI (1994a) Dynamic models of mass-dependant predation, risk-sensitive foraging, and premigratory fattening in birds. Ecology 75:1131–1140

Bednekoff PA, Houston AI (1994b) Avian daily foraging patterns: effects of digestive constrains and variability. Evol Ecol 8:36–52

Bednekoff PA, Krebs J (1995) Great Tit fat reserves: effects of changing and unpredictable feeding day length. Funct Ecol 9:457–462

Bednekoff PA, Biebach H, Krebs J (1994) Great Tit fat reserves under unpredictable temperatures. J Avian Biol 25:156–160

Beecher MD, Beecher IM (1979) Sociobiology of bank swallows: reproductive strategy of the male. Science 205:1282–1285

Biebach H (1998) Phenotypic organ flexibility in Garden Warblers *Sylvia borin* during long-distance migration. J Avian Biol 29:529–535

Blem CR (1975) Geographic variation in wing-loading of the House Sparrow. Wilson Bull 87:543–549

Blem CR (1976) Patterns of lipid storage and utilization in birds. Am Zool 16:671–684

Brodin A (2000) Why do hoarding birds gain fat in winter in the wrong way? Suggestions from a dynamic model. Behav Ecol 11:27–39

Burns JG, Ydenberg RC (2002) The effects of wing loading and gender on the escape flights of least sandpipers (*Calidris minutilla*) and western sandpipers (*Calidris mauri*). Behav Ecol Sociobiol 52:128–136

Chai P (1997) Hummingbird hovering energetics during moult of primary flight feathers. J Theor Biol 200:1527–1536

Chai P, Dudley R (1999) Maximum flight performance of hummingbirds: capacities, constraints, and trade-offs. Am Nat 153(4):398–411

Chai P, Altshuler DL, Stephens DB, Dillon ME (1999) Maximal horizontal flight performance of hummingbirds: effects of body mass and molt. Physiol Biochem Zool 72(2):145–155

Cimprich DA, Woodrey MS, Moore FR (2005) Passerine migrants respond to variation in predation risk during stopover. Anim Behav 69:1173–1179

Clark CW, Ekman J (1995) Dominant and subordinate fattening strategies: a dynamic game. Oikos 72:205–212

Cresswell W (1993) Escape responses by Redshanks, *Tringa totanus*, on attack by avian predators. Anim Behav 46:609–611

Cresswell W (1996) Surprise as a winter hunting strategy in Sparrowhawks *Accipiter nisus*, Peregrines *Falco peregrinus* and Merlins *F. columbarius*. Ibis 138:684–692

Cresswell W, Clark JA, Macleod R (2009) How climate change might influence the starvation-predation risk trade-off response. Proc R Soc Lond B 276:3553–3560

Dall SRX, Witter MS (1998) Feeding interruptions, diurnal mass change and daily routines of behaviour in the Zebra Finch. Anim Behav 55:715–725

Davidson NC, Evans PR (1988) Prebreeding accumulation of fat and muscle protein by arctic-breeding shorebirds. In: Oullet H (ed) Acta XIX Congressus Internationalis Ornithologici, Ottawa, ON 1986. Natural Museum of Natural Sciences, Ottawa

Davies NB (2000) Cuckoos, cowbirds and other cheats. Academic, London

Dawkins R, Krebs JR (1979) Arms races between and within species. Proc R Soc Lond B 205:489–511

Dolnik VR, Bluymental TI (1967) Autumnal premigratory and migratory periods in the Chaffinch (*Fringilla coelebs coelebs*) and some other temperate-zone birds. Condor 69:435–468

Dolnik VR, Gavrilov VM (1979) Bioenergetics of molt in the Chaffinch (*Fringilla coelebs*). Auk 96:253–264

Ekman JB, Hake MK (1990) Monitoring starvation risk: adjustments of body reserves in Greenfinches (*Carduelis chloris* L.) during periods of unpredictable foraging success. Behav Ecol 1:62–67

Flinks H, Kolb H (1997) Body mass variations of Stonechats *Saxicola torquata* during breeding and moult. Vogelwelt 118(1):1–10

Francis CM, Wood DS (1989) Effects of age and wear on wing length of Wood-Warblers. J Field Ornithol 60:495–503

Fransson T (1995) Timing and speed of migration in North and West European populations of *Sylvia* warblers. J Avian Biol 26:39–48

Fransson T (1997) Time and energy in long-distance bird migration. Ph.D. thesis, Stockholm University, Stockholm

Fransson T, Jakobsson S (1998) Fat storage in male Willow Warblers in spring: do residents arrive lean or fat? Auk 115:759–763

Fransson T, Weber T (1997) Migratory fuelling in Blackcaps (*Sylvia atricapilla*) under perceived risk of predation. Behav Ecol Sociobiol 41:75–80

Fry CH, Ash JS, Ferguson-Lees IJ (1970) Spring weights of some Palaearctic migrants at Lake Chad. Ibis 112:58–82

Fry CH, Ferguson-Lees IJ, Dowsett RJ (1972) Flight muscle hypertrophy and ecophysiological variation of Yellow Wagtail *Motacilla flava* races at Lake Chad. J Zool Lond 167:293–306

Gaunt AS, Hikida RS, Jehl JR, Fenbert L (1990) Rapid atrophy and hypertrophy of an avian flight muscle. Auk 107:649–659

Gessaman JA, Workman GW, Fuller MR (1991) Flight performance, energetics and water turn-over of tippler pigeons with a harness and dorsal load. Condor 93:546–554

Ginn HB, Melville DS (1983) Moult in birds, BTO guide no 19. British Trust for Ornithology, Tring

Gosler AG (1996) Environmental and social determinants of winter fat storage in the Great Tit *Parus major*. J Anim Ecol 65:1–17

Gosler AG, Greenwood JJD, Perrins C (1995) Predation risk and the cost of being fat. Nat Lond 377:621–623

Götmark F, Post P (1996) Prey selection by Sparrowhawks, *Accipiter nisus*: relative predation risk for breeding passerine birds in relation to their size, ecology and behaviour. Philos Trans R Soc Lond B 351:1559–1577

Grubb TC Jr, Pravosudov VV (1994) Towards a general theory of energy management in wintering birds. J Avian Biol 25:255–260

Haftorn S (1972) Hypothermia of tits in the arctic winter. Ornis Scand 3:153–166

Haftorn S (1989) Seasonal and diurnal body weight variations in titmice, based on analyses of individual birds. Wilson Bull 101:217–235

Haukioja E (1971) Flightlessness in some moulting passerines in Northern Europe. Ornis Fenn 48:101–116

Hedenström A (1992) Flight performance in relation to fuel loads in birds. J Theor Biol 158:535–537

Hedenström A, Alerstam T (1995) Optimal flight speed of birds. Philos Trans R Soc Lond B 348:471–487

Hedenström A, Sunada S (1999) On the aerodynamics of moult gaps in birds. J Exp Biol 202(1):67–76

Hedenström A, Rosén M (2001) Predator versus prey: on aerial hunting and escape strategies in birds. Behav Ecol 12:150–156

Houston AI, McNamara JM (1993) A theoretical investigation of the fat reserves and mortality levels of small birds in winter. Ornis Scand 24:205–219

Houston AI, McNamara JM, Hutchinson JMC (1993) General results concerning the trade-off between gaining energy and avoiding predation. Philos Trans R Soc Lond B 341:375–397

Houston DC, Donnan D, Jones P (1995a) The sources of nutrients required for egg production in Zebra Finches *Poephila guttata*. J Zool Lond 235:469–483

Houston DC, Donnan D, Jones P, Hamilton I, Osborne D (1995b) Changes in muscle condition of female Zebra Finches *Poephila guttata* during egg laying and the role of protein storage in bird skeletal muscle. Ibis 137:322–328

Houston DC, Donnan D, Jones P (1995c) Use of labelled methionine to investigate the contribution of muscle proteins to egg production in Zebra Finches. J Comp Physiol B 164:161–164

Houston AI, Welton NJ, McNamara JM (1997) Acquisition and maintenance costs in the long-term regulation of avian reserves. Oikos 78:331–340

Howland HC (1974) Optimal strategies for predator avoidance: the relative importance of speed and manoeuvrability. J Theor Biol 47:333–350

Hume ID, Biebach H (1996) Digestive tract function in the long-distance migratory Garden Warbler, *Sylvia borin*. J Comp Physiol B 166:388–395

Jehl JR (1997) Cyclical changes in body composition in the annual cycle and migration of the Eared Grebe *Podiceps nigricollis*. J Avian Biol 28(2):132–142

Jenni L, Jenni-Eiermann S (1998) Fuel supply and metabolic constraints in migrating birds. Avian Biol 29:521–528

Jenni L, Winkler L (1994) Moult and ageing of European passerines. Academic, London

Jones G (1986) Sexual chases in Sand Martins (*Riparia riparia*): cues for males to increase their reproductive success. Behav Ecol Sociobiol 19:179–185

Jones MM (1990) Muscle protein loss in laying House Sparrows *Passer domesticus*. Ibis 133:193–198

Jones PJ, Ward P (1976) The level of reserve protein as the proximate factor controlling the timing of breeding and clutch size in the Red-billed Quelea. Ibis 118:547–574

Kendall MD, Ward P, Bacchus S (1973) A protein reserve in the pectoralis major muscle of *Quela quela*. Ibis 115:600–601

Kenward RE (1978) Hawks and doves: factors affecting success and selection in Goshawk attacks on Woodpigeons. J Anim Ecol 47:449–460

Klaassen M, Biebach H (1994) Energetics of fattening and starvation in the long-distance migratory Garden Warbler, *Sylvia borin*, during the migratory phase. J Comp Physiol B 164:362–371

Krebs JR (1970) The efficiency of courtship feeding in the Blue Tit, *Parus caeruleus*. Ibis 112:108–110

Kullberg C (1998) Does diurnal variation in body mass affect take-off ability in wintering willow tits? Anim Behav 56:227–233

Kullberg C, Fransson T, Jakobsson S (1996) Impaired predator evasion in fat Blackcaps. Proc R Soc Lond B 263:1671–1675

Kullberg C, Jakobsson S, Fransson T (1998) Predator induced take-off strategy in Great Tits (*Parus major*). Proc R Soc Lond B 265:1659–1664

Kullberg C, Jakobsson S, Fransson T (2000) High migratory fuel load impair predator evasion in Sedge Warblers. Auk 117:1034–1038

Kullberg C, Houston DC, Metcalfe NB (2002a) Impaired flight ability –a cost of reproduction in female blue tits. Behav Ecol 13:575–579

Kullberg C, Metcalfe NB, Houston DC (2002b) Impaired flight ability during incubation in the pied flycatcher. J Avian Biol 33:179–183

Kullberg C, Jakobsson S, Kaby U, Lind J (2005) Impaired flight ability prior to egg-laying: a cost of being a capital breeder. Funct Ecol 19:98–101

Lee SJ, Witter MS, Cuthill IC, Goldsmith AR (1996) Reduction in escape performance as a cost of reproduction in gravid Starlings, *Sturnus vulgaris*. Proc R Soc Lond B 263:619–624

Lehikoinen E (1987) Seasonality of the daily weight cycle in wintering passerines and its consequences. Ornis Scand 18:216–226

Lilliendahl K (1997) The effect of predator presence on body mass in captive Greenfinches. Anim Behav 53:75–81

Lilliendahl K (1998) Yellowhammers get fatter in the presence of a predator. Anim Behav 55:1335–1340

Lilliendahl K, Carlson A, Welander J, Ekman JB (1996) Behavioural control of daily fattening Great Tits (*Parus major*). Can J Zool 74:1612–1616

Lima SL (1985) Collective detection of predatory attack by social foragers – fraught with ambiguity. Anim Behav 50:1097–1108

Lima SL (1986) Predation risk and unpredictable feeding conditions: determinants of body mass in birds. Ecology 67:377–385

Lima SL (1993) Ecological and evolutionary perspectives on escape from predatory attack: a survey of North American birds. Wilson Bull 105:1–215

Lima SL, Dill LM (1990) Behavioral decisions made under the risk of predation: a review and prospectus. Can J Zool 68:619–640

Lind J (2001) Escape flight in moulting Tree Sparrows (*Passer montanus*). Funct Ecol 15:29–35

Lind J (2004) What determines probability of surviving predator attacks in bird migration? The relative importance of vigilance and fuel load. J Theor Biol 231:223–232

Lind J, Cresswell W (2005) Determining the fitness consequences of anti-predation behavior. Behav Ecol 16:945–956

Lind J, Cresswell W (2006) Anti-predation behaviour during bird migration; the benefit of studying multiple behavioural dimensions. J Ornithol 147:310–316

Lind J, Jakobsson S (2001) Body building and concurrent mass loss: flight adaptations in tree sparrows. Proc R Soc Lond B 268:1915–1919

Lind J, Fransson T, Jakobsson S, Kullberg C (1999) Reduced take-off ability in robins due to migratory fuel load. Behav Ecol Sociobiol 46(1):65–70

Lind J, Gustin M, Sorace A (2004) Compensatory bodily changes during moult in Tree Sparrows Passer montanus in Italy. Ornis Fenn 81:75–83

Lindström Å (1989) Finch flock size and risk of hawk predation at a migratory stopover site. Auk 106:225–232

Lindström Å, Daan S, Visser GH (1994) The conflict between moult and migratory fat deposition: a photoperiodic experiment with Bluethroats. Anim Behav 48:1173–1181

Lindström Å, Kvist A, Piersma T, Dekinga A, Dietz MW (2000) Avian pectoral muscle size rapidly tracks body mass changes during flight, fasting and fuelling. J Exp Biol 203:913–919

Lindström Å, Piersma T (1993) Mass changes in migrating birds: the evidence for fat and protein storage re-examined. Ibis, 135:70–78

MacLeod R, Lind J, Clark J, Cresswell W (2007) Mass regulation in response to predation risk can indicate population declines. Ecol Lett 10:945–955

Magnhagen C (1991) Predation risk as a cost of reproduction. Trends Ecol Evol 6(6):183–185

Marden JH (1987) Maximum lift production during takeoff in flying animals. J Exp Biol 130:235–258

Marsh RL (1984) Adaptations of the Gray catbird *Dumetella carolinensis* to long-distance migration: flight muscle hypertrophy associated with elevated body mass. Physiol Zool 57(1):105–117

McLandress MR, Raveling DG (1981) Changes in diet and body composition of Canada geese before spring migration. Auk 98:65–79

McNamara JM, Houston AI (1990) The value of fat reserves and the trade-off between starvation and predation. Acta Biotheor 38:37–62

McNamara JM, Houston AI, Krebs JR (1990) Why hoard? The economics of food storing in tits, *Parus* spp. Behav Ecol 1:12–23

McNamara JM, Houston AI, Lima SL (1994) Foraging routines of small birds in winter: a theoretical investigation. J Avian Biol 25:287–302

Meijer T, Drent R (1999) Re-examination of the capital and income dichotomy in breeding birds. Ibis 141:399–414

Metcalfe NB, Furness RW (1984) Changes priorities: the effect of pre-migratory fattening on the trade-off between foraging and vigilance. Behav Ecol Sociobiol 15:203–206

Metcalfe NB, Ure SE (1995) Diurnal variation in flight performance and hence potential predation risk in small birds. Proc R Soc Lond B 261:395–400

Monaghan P, Nager RG (1997) Why don't birds lay more eggs? Trends Ecol Evol 12(7):270–274

Monaghan P, Nager RG, Houston DC (1998) The price of eggs: increased investment in egg production reduces the offspring rearing capacity of parents. Proc R Soc Lond B 265:1731–1735

Moore FR (1994) Resumption of feeding under risk of predation: effect of migratory condition. Anim Behav 48:975–977

Moore FR, Simms PA (1986) Risk-sensitive foraging by a migratory bird (*Dendroica coronata*). Experientia 42:1054–1056

Moreno J (1989) Strategies of mass change in breeding birds. Biol J Linn Soc 37:297–310

Naef-Daenzer B (1993) A new transmitter for small animals and enhanced methods of home-range analysis. J Wildl Manage 57:680–689

Nakamura M (1990) Cloacal protuberance and copulatory behavior of the Alpine Accentor (*Prunella collaris*). Auk 107:284–295

Newton I (1966) The moult of the Bullfinch *Pyrrhula pyrrhula*. Ibis 108:41–67

Newton I (1986) The sparrowhawk. Poyser, Carlton

Norberg Å (1981) Temporary mass decrease in breeding birds may result in more fledged young. Am Nat 118:838–850

Norberg UM (1990) Vertebrate flight. Springer, Berlin

Norberg UM (1995) How a long tail and changes in mass and wing shape affect the cost for flight in animals. Funct Ecol 9:48–54

Pennycuick CJ (1969) The mechanics of bird migration. Ibis 111:525–556

Pennycuick CJ (1975) Mechanics of flight. In: Farner DS, King JR (eds) Avian biology, vol 5. Academic, New York, pp 1–73

Pennycuick CJ (1989) Bird flight performance. A practical calculation manual. Oxford University Press, Oxford

Perrins C (1996) Eggs, egg formation and the timing of breeding. Ibis 138:2–15

Piersma T (1998) Phenotypic plasticity during migration: optimization of organ size contingent on the risks and rewards of fueling and flight? J Avian Biol 29:511–520

Piersma T, Lindström Å (1997) Rapid reversible changes in organ size as a component of adaptive behaviour. Trends Ecol Evol 12(4):134–138

Pravosudov VV, Grubb TC (1998) Management of fat reserves in tufted titmice *Baelophus bicolor* in relation to risk of predation. Anim Behav 56:49–54

Rappole JH, Tipton AR (1991) New harness design for attachment of radiotransmitters to small passerines. J Field Ornithol 62:335–337

Rayner JMV (1988) Form and function in avian flight. Curr Ornithol 5:1–66

Reinertsen RE, Haftorn S (1984) The effect of short-time fasting on metabolism and nocturnal hypothermia in the Willow Tit *Parus montanus*. J Comp Physiol B 154:23–28

Rogers CM (1987) Predation risk and fasting capacity: do wintering birds maintain optimal body mass? Ecology 68:1051–1061

Rothstein SI, Verner J, Stevens E (1984) Radio-tracking confirms a unique diurnal pattern of spatial occurrence in the parasitic Brown-headed Cowbird. Ecology 65:77–88

Rudebeck G (1950) The choice of prey and modes of hunting of predatory birds with special reference to their selective effect. Oikos 2:65–88

Savile DBO (1957) Adaptive evolution in the avian wing. Evolution 11:212–224

Schmidt-Nielsen K (1990) Animal psysiology: adaptation and environment. 4th edn Cambridge University Press, Cambridge

Scott DM, Ankney CD (1983) The laying cycle of Brown-headed Cowbirds: passerine chickens? Auk 100:583–592

Selman RG, Houston DC (1996) A technique for measuring lean pectoral muscle mass in live small birds. Ibis 138:348–350

Slagsvold T (1982) Sex, size and natural selection in the Hooded Crow Corvus corone cornix. Ornis Scand 13:165–175

Slagsvold T, Dale S (1996) Disappearance of female Pied Flycatchers in relation to breeding stage and experimentally induced molt. Ecology 77:461–471

Solonen T (1997) Effect of Sparrowhawk *Accipiter nisus* predation on forest birds in southern Finland. Ornis Fenn 74:1–14

Srygley RB, Dudley R (1993) Correlations of the position of the center of body-mass with butterfly escape tactics. J Exp Biol 174:155–166

Stearns SC (1992) The evolution of life histories. Oxford University Press, Oxford

Swaddle JP, Biewener AA (2000) Exercise and reduced muscle mass in Starlings. Nat Lond 406:585–586

Swaddle JP, Witter MS (1997) The effects of moult on the flight performance, body mass, and behavior of the European Starlings (*Sturnus vulgaris*): an experimental approach. Can J Zool 75:1135–1146

Swaddle JP, Witter MS, Cuthill IC, Budden A, McCowen P (1996) Plumage condition affects flight performance in common Starlings: implications for developmental homeostasis, abrasion and moult. J Avian Biol 27:103–111

van den Hout PJ, Piersma T, Dekinga A, Lubbe SK, Visser GH (2006) Ruddy turnstones *Arenaria interpres* rapidly build pectoral muscle after raptor scares. J Avian Biol 37(5):425–430

van der Veen IT (1999) Effects of predation risk on diurnal mass dynamics and foraging routines of Yellowhammers (*Emberiza citrinella*). Behav Ecol 10:545–551

van der Veen IT, Lindström KM (2000) Escape flights of Yellowhammers and Greenfinches: more than just physics. Anim Behav 59:593–601

Veasey JS (1999) Egg production, flight velocity and predation risk in birds. PhD-thesis, Glasgow University, Glasgow

Veasey JS, Metcalfe NB, Houston DC (1998) A reassessment of the effect of body mass upon flight speed and predation risk in birds. Anim Behav 56:883–889

Veasey JS, Houston DC, Metcalfe NB (2000a) Flight muscle atrophy and predation risk in breeding birds. Funct Ecol 14:115–121

Veasey JS, Houston DC, Metcalfe NB (2000b) A hidden cost of reproduction: the trade-off between clutch size and escape take-off speed in female Zebra Finches. J Anim Ecol 70:20–24

Ward P (1969) The annual cycle of the Yellow-vented Bulbul *Pycnonotus goiavier* in a humid equatorial environment. J Zool Lond 157:25–45

Witter MS, Cuthill IC (1993) The ecological cost of avian fat storage. Philos Trans R Soc Lond B 340:73–92

Witter MS, Cuthill IC, Bonser RH (1994) Experimental investigations of mass-dependent predation risk in the European Starling, *Sturnus vulgaris*. Anim Behav 48:201–222

Witter MS, Swaddle JP, Cuthill IC (1995) Periodic food availability and strategic regulation of body mass in the European Starling, *Sturnus vulgaris*. Funct Ecol 9:568–574

Woodburn RJW, Perrins CM (1997) Weight changes and the body reserves of female Blue Tits, *Parus caeruleus*, during the breeding season. J Zool Lond 243:789–802

Chapter 2
Dietary Calcium Availability and Reproduction in Birds

S. James Reynolds and Christopher M. Perrins

The total amount of calcium circulating in the blood of an average hen at any one time is about 25 milligrams. Hence an amount of calcium equal to the weight of calcium present in the circulation is removed from the blood every 12 minutes during the main period of shell calcification. Where does this calcium come from?

Taylor (1970)

2.1 Introduction

Reproduction is energetically costly. Egg formation in birds requires 37–55% of basal metabolism for small passerines and 160–216% for ducks and the Southern Brown Kiwi (*Apteryx australis*) (Walsberg 1983). The production of eggs requires an adequate supply of water, macronutrients (carbohydrate, protein, and fat), and micronutrients (essential fatty acids, amino acids, vitamins, and ions). Of this latter group of nutrients, calcium is probably the most limiting micronutrient required by the laying bird (Burley and Vadehra 1989); 98% of the dry mass of the eggshell consists of calcite, a crystalline form of calcium carbonate ($CaCO_3$) (Romanoff and Romanoff 1949). The eggshell encases all of the nutrients required for embryonic development and survival during the incubation period, preventing the developing chick from being crushed by the incubating adult. The shell also protects the embryo from dehydration, prevents the entry of pathogens from the external environment, and allows gaseous exchange. It also acts as a source of calcium for skeletal development of the embryo. The withdrawal from the skeleton results in

S.J. Reynolds (✉)
Centre for Ornithology, School of Biosciences, College of Life & Environmental Sciences, University of Birmingham, Edgbaston, Birmingham, B15 2TT, UK
e-mail: J.Reynolds.2@bham.ac.uk

C.F. Thompson (ed.), *Current Ornithology Volume 17*, Current Ornithology, DOI 10.1007/978-1-4419-6421-2_2, © Springer Science+Business Media, LLC 2010 "natural thinning" of the eggshell from ~6% (e.g., Balkan et al. 2006) to 21% (e.g., Booth and Seymour 1987) as the embryo absorbs calcium from the shell during incubation. After egg production, the calcium requirements of the adult in many species must remain elevated during chick-rearing, when birds feed calcium-rich foods to their young (see Table 3 in Graveland 1996a) for the continued mineralization of their skeletons (Starck 1998).

In stark contrast to our detailed knowledge of the macronutrients, little is known about the micronutrient requirements for successful avian reproduction beyond isolated studies on trace elements and their importance during egg laying in poultry (e.g., copper, Baumgartner et al. 1978; selenium, Jensen 1968; zinc, Savage 1968). Many small passerines produce large clutches of eggs, requiring considerable nutritional investment (macro- and micronutrients) for the production of viable eggs. For example, the mass of a Blue Tit (*Cyanistes caeruleus*) clutch may represent 130% or more of the female's body mass (Perrins and Birkhead 1983). Although clutch formation requires a large investment of lipid (e.g., Schifferli 1980) and protein (Robbins 1981), micronutrient requirements are also considerable; the clutches of small passerines often contain more calcium than is present in the entire skeleton of the breeding female (Graveland 1995).

Unlike macronutrients, some of which may be laid down as deposits weeks or months before laying (e.g., Korschgen 1977), calcium is not stored to any great extent by birds during the prelaying period (Pahl et al. 1997; but see Piersma et al. 1996). For those species that do not routinely consume calcium-rich foods, females increase their intake of dietary calcium just before the onset of egg laying (e.g., Great Tit [*Parus major*], Graveland and Berends 1997).

Although an adequate supply of dietary calcium is fundamental to successful egg formation and chick development, surprisingly little is known about calcium uptake and utilization by breeding birds, other than by the Domestic Chicken (*Gallus domesticus*). By comparison, similar information for the macronutrient requirements of breeding birds is detailed. Macronutrients are obtained either from endogenous reserves or from the diet, or both (Meijer and Drent 1999; Drent and Daan 1980). The principal macronutrients in avian reproduction are fat and protein. Birds use a range of strategies to obtain egg nutrients. The American Coot (*Fulica americana*) is not constrained by endogenous reserves but instead accumulates fat reserves during egg laying (Arnold and Ankney 1997), whereas the Common Eider (*Somateria mollissima*) spends the prelaying period foraging and then virtually stops feeding at the onset of egg laying (Korschgen 1977). In many species, protein reserves decline during egg production (e.g., Adélie Penguin [*Pygoscelis adeliae*], Astheimer and Grau 1985; Red-billed Quelea [*Quelea quelea*], Jones and Ward 1976), but in others depletion does not occur (e.g., Brown-headed Cowbird [*Molothrus ater*], Ankney and Scott 1980; Wood Duck [*Aix sponsa*], Drobney 1980). In the Glossy Swiftlet (*Collocalia esculenta*), daily intake of protein is three to four times the daily requirement for clutch formation (Hails and Turner 1985).

In some species, lipid reserves are depleted for egg formation (e.g., Ring-necked Duck [*Aythya collaris*], Hohman 1986; Mallard [*Anas platyrhynchos*], Krapu 1981), but protein is supplied almost exclusively from foraging. In other species, such as the Wood Duck, protein and lipid are inextricably linked. Although protein for egg formation is derived from the diet and somatic protein reserves are not depleted, lipid reserves provide 88% of the energetic requirements for the biosynthesis of nonprotein fractions (Drobney 1980).

Whatever the strategy of macronutrient provisioning during egg laying in birds, accounts of wholesale macronutrient-limited reproduction are absent from the avian literature. Indeed, some species appear to experience few problems meeting their macronutrient requirements for clutch formation (e.g., Nager et al. 1997).

In this chapter, we review the literature on dietary calcium in light of the growing concern about its declining availability in certain ecosystems. Rates of acidification in some aquatic and terrestrial ecosystems have increased dramatically in the last few decades, resulting in the loss of cations such as calcium from upper soil horizons and their replacement by toxic cations such as aluminum, cadmium, and lead (van Breemen et al. 1983). The adverse effects of acidification on commercial fish stocks (e.g., Haines and Baker 1986) and on timber (e.g., DesGranges et al. 1987) have been appreciated for some time, but effects on avian populations have not been considered until relatively recently (see review by Graveland 1998).

This chapter reviews the literature on incidents of calcium-specific foraging behavior to show that the significant increase in dietary calcium intake, prior to the onset of egg laying, is probably a widespread phenomenon occurring in all avian species that do not routinely eat a calcium-rich diet. We also discuss calcium-rich food items fed to chicks. We review current knowledge concerning the utilization of dietary calcium during egg production in the Domestic Chicken and in other avian species. We suggest that while the poultry industry has invested considerable research effort in improving the composition of calcium-rich diets for laying birds (see reviews by Etches 1987; Petersen 1965), relatively little is known about the dietary calcium requirements for noncommercial species. Indeed, evidence of calcium-limited reproduction in free-living birds was provided only relatively recently by Graveland (1995). We also consider the effects of limited calcium availability during egg formation on eggshell structure and chick development as well as on adult physiology and behavior. We discuss how close some birds may be to a dietary calcium threshold, below which their breeding performance might decline severely. Finally, we evaluate the effectiveness of measures such as calcium supplementation in increasing dietary calcium availability and we suggest further research to increase our understanding of the calcium requirements of breeding birds.

2.2 Calcium Requirements for Successful Breeding

2.2.1 Calcium-Specific Foraging During Egg Laying

The extent of the dietary shift by birds to more calcium-rich foods during breeding depends upon the nutritional strategy of the species in question. Many birds consume calcium-rich foods on a daily basis during egg production. Table 2.1 summarizes

Table 2.1 Calcium-rich materials consumed by wild egg laying birds

Material	Species	References
Snail shell	Pigeons (*Columba* spp.) and doves (*Streptopelia* spp.)	Murton et al. 1964
	Great Tit (*Parus major*)	Graveland 1996a
	Coal Tit (*Periparus ater*)	Betts 1955
	Red-billed Quelea (*Quelea quelea*)	Jones 1976
	House Sparrow (*Passer domesticus*)	Schifferli 1976
	Brown-headed Cowbird (*Molothrus ater*)	Ankney and Scott 1980
	Spotted Flycatcher (*Muscicapa striata*)	Davies 1977
	European Pied Flycatcher (*Ficedula hypoleuca*)	Eeva and Lehikoinen 1995
	Black-throated Blue Warbler (*Dendroica caerulescens*)	Taliaferro et al. 2001
	Wild Turkey (*Meleagris gallopavo*)	Beasom and Pattee 1978
	Goldcrest (*Regulus regulus*) and Firecrest (*R. ignicapilla*)	Thaler 1979
Vertebrate bones	European Golden Plover (*Pluvialis apricaria*)	Byrkjedal 1975
	Sandpipers (*Calidris* spp.)	MacLean 1974
	Red-cockaded Woodpecker (*Picoides borealis*)	Repasky et al. 1991
	Red Crossbill (*Loxia curvirostra*)	Payne 1972
Grit	Common Pheasant (*Phasianus colchicus*)	Kopischke 1966; Kopischke and Nelson 1966; Korschgen 1964; Harper 1964; Sadler 1961
	Band-tailed Pigeon (*Columba fasciata*)	March and Sadleir 1975
	Red-billed Quelea	Jones 1976
	House Sparrow	Pinowksa and Kraśnicki 1985a
Chicken eggshell	Great Tit	Graveland 1996a
Mortar between bricks	Red Crossbill	Sušic 1981
White-wash	Pearl-breasted Swallow (*Hirundo dimidiata*)	Dean 1989
Ash	Boreal Chickadee (*Parus hudsonicus*)	Ficken 1989
	Rufous Hummingbird (*Selasphorus rufus*)	Adam and des Lauriers 1998
Soil	Dwarf Cassowary (*Casuarius bennetti*)	Symes et al. 2006
Decaying wood	Parrot Crossbill (*Loxia pytyopsittacus*) and Common Redpoll (*Carduelis flammea*)	Pulliainen et al. 1978

observations of calcium-specific foraging in birds during egg laying. Although carnivorous birds such as vultures include bone fragments in their regular diet, the calcium content of the diet is low (e.g., 0.12% for the Cape Vulture [*Gyps coprotheres*], Mundy et al. 1992), as is their digestive efficiency (e.g., 14% for vultures [*Gyps* spp.] that consume a 0.5 mg Ca per gram of meat diet, Houston 1978). Insectivores and nectarivores typically consume low calcium diets, which often leads them to supplement their diets with additional calcium-rich items. Insectivores supplement their diets with fragments of bone, snail shell, and eggshell (e.g., Spotted Flycatcher [*Muscicapa striata*], Davies 1977), and they increase grit ingestion to aid their digestive processing of hard-bodied insects, such as Coleoptera, in the gizzard. In comparison with carnivores and insectivores, gizzards of nectarivores are relatively small (e.g., Karasov et al. 1986) because they do not consume hard food items. However, hummingbirds eat tiny fragments of hard, calcium-rich material (e.g., Verbeek 1971), which they beak-stab. One-third of all avian species in tropical and temperate communities eat predominantly fruit (Klasing 1998). While most fruits (except figs; see O'Brien et al. 1998) eaten by birds are low in calcium, some fruiting trees may be important stopover sites for migrating birds providing them with important sources of dietary calcium while en route to breeding grounds (e.g., Foster 2007). Frugivores have short intestines with large lumens, but the gizzard is small (e.g., Walsberg 1975). A fruit-only diet can result in deficiencies of certain amino acids and electrolytes (Studier *et al.* 1988), so many frugivores eat snails and grit in order to restore nutrient balances (e.g., White-crowned Pigeon [*Patagioenas leucocephala*], Wiley and Wiley 1979).

Calcium supplementation of breeding diets is also essential for birds whose diet is made up predominantly of plant material other than fruits. For example, the Spruce Grouse (*Falcipennis canadensis*) consumes spring foods (flowers of Trailing Arbutus [*Epigaea repens*] and capsules of moss *Polytrichum* spp.) that provide only ~45% of the calcium required for clutch formation (Naylor and Bendell 1989). Bendell-Young and Bendell (1999) suggest that calcareous grit provides an additional source of dietary calcium. Other granivores, such as the Zebra Finch (*Taeniopygia guttata*), also supplement their low calcium seed diets with calcium-rich material (Zann and Straw 1984). In addition to providing dietary calcium, ingested calcareous material may facilitate the crushing of seeds by the ventral and dorsal grinding plates of the gizzard (see Gionfriddo and Best 1999). The harder the food items consumed, the greater the grit content of the gizzard. Grit is consumed by many birds (see Diamond et al. 1999; Gionfriddo and Best 1999), but whether it simply aids the grinding action of the gizzard (Gionfriddo and Best 1996) or provides calcium to the laying bird (Bendell-Young and Bendell 1999) is sometimes difficult to resolve. Many birds consume a range of hard materials (e.g., bones, insects, seeds, teeth) that are retained for a long time in the gizzard and function as grit substitutes (see Table IV in Gionfriddo and Best 1999). Clearly, hard seeds are retained solely to enhance the efficiency of mechanical grinding by the gizzard and not as a source of dietary calcium.

In omnivores, the extent of dietary shifts to meet the extra nutritional demands of reproduction depends upon the quantities of calcium-rich foods that are routinely eaten at other times of the year. Within a single population of dietary generalists, birds can show significantly different breeding performances according to their dietary specializations during the breeding season. For example, Pierotti and Annett (1991) reported that Herring Gulls (*Larus argentatus*) on Great Island, Newfoundland, that fed on Blue Mussels (*Mytilus edulis*), produced significantly larger and heavier clutches and hatched more eggs than birds feeding on garbage, Leach's Storm

Petrels (*Oceanodroma leucorhoa*), or more general diets. They also laid their eggs earlier than other birds (Pierotti and Annett 1987) and were the only birds to produce seven or more eggs with no gaps in laying during an egg-removal experiment. Mussel-feeding specialists were probably not as calcium-stressed during egg laying as were petrel-feeding and garbage-feeding specialists because, although low in energy and protein, mussels are rich in calcium and manganese. Gulls that consumed garbage and petrels obtained calcium predominantly from bone material, in which calcium is relatively insoluble, sequestered in a crystalline lattice of apatite. In contrast, mussels have a high content of easily metabolized calcium, and their consumption during egg formation may have alleviated calcium stress in laying females. Furthermore, early skeletal development of embryos depends upon the availability of sulfonated amino acids that are found in abundance in mussels. Garbage consists of connective tissue that is low in such amino acids. Sibly and McCleery (1983) found that only female Herring Gulls at Walney Island, UK, consumed mussels during laying, while their mates continued to specialize on garbage, suggesting that females were probably obtaining calcium for reproduction from ingested mussels.

2.2.2   Calcium-Specific Foraging During Chick-Rearing

After hatching, a chick undergoes rapid growth during which its anatomy develops and tissues mature, resulting in large increases in overall body mass. Postnatal growth is one of the most energy-demanding periods of a bird's life, and in many species parental provisioning plays a vital role in the growth and long-term survival of their offspring. In some birds (e.g., *Larus* spp.), adults show discrete shifts in foraging behavior in order to provide a full complement of nutrients for their growing chicks. For example, Western Gulls (*Larus occidentalis*) nesting on Alcatraz Island, USA, switched from predominantly nutrient-poor garbage to nutrient-rich Northern Anchovies (*Engraulis mordax*) after their chicks hatched (Annett and Pierotti 1989). Furthermore, birds prey-switched again, this time to the larger Plainfin Midshipman (*Porichthys notatus*) when their chicks were 2–3 weeks old. There is some evidence to suggest that even within a single population, the growth of chicks can vary considerably with the food items that their parents provide (e.g., Pigeon Guillemot [*Cepphus columba*], Golet et al. 2000).

The calcium demands of developing chicks increase rapidly over the first 2 weeks posthatching. Precocial young are capable of feeding themselves almost as soon as they hatch. In *Calidris* sandpipers, chicks leave the nest within a day of hatching and are led by adult birds to feeding sites, but adults do not feed the young directly. MacLean (1974) found bones and teeth of Brown Lemmings (*Lemus trimucronatus*) in the stomachs of 22.2% of juvenile Dunlins (*Calidris alpina*) and 14.6% of juvenile Pectoral Sandpipers (*Calidris melanotos*) examined in July, when most growth occurs, but in only 0 and 2.7%, respectively, of birds examined in August. In precocial Galliformes, such as the Common Pheasant (*Phasianus colchicus*) and the Ruffed Grouse (*Bonasa umbellus*), grit ingestion by chicks may occur within a day of hatching (Dalke 1938 and Bump et al. 1947, respectively).

Altricial young require adults to deliver food to the nest before they fledge. Bilby and Widdowson (1971) found that the calcium content of tissues of chicks increased by seven to eight times from between hatching and 11–12 days of age for Common Blackbirds (*Turdus merula*) (from 70 to 510 mg per 100 g fat-free body mass) and for Song Thrushes (*T. philomelos*) (from 60 to 450 mg). Over this period, the mean body mass of Blackbird nestlings increased 14 times and total somatic calcium content increased by ~100 times (Fig. 2.1). At hatching, femora contained ~1% calcium and were predominantly cartilaginous. However, by the time the chicks were ready to leave the nest approximately 2 weeks later, long bones had developed calcified cortical (structural) bone deposits. Bilby and Widdowson (1971) also found that the gut contents of nestling Common Blackbirds and Song Thrushes sometimes contained as much as half the total calcium contained in their bodies. Although many of their food items are apparently nutrient-poor, nestlings also ingest earthworms (*Lumbricus* spp.), caterpillars, and insects, all of which excrete calcium as a primary excretory product. In addition, earthworms and caterpillars consume calcium-rich soil and leaves, respectively.

In other species, adult birds supplement the diet of their young with calcium-rich material. Fragments of fish bone, crayfish (*Orconectes virilis*) exoskeleton, and eggshells of Herring Gulls and Great Northern Loons (*Gavia immer*) have been found in the nestboxes and in the stomachs of nestling Tree Swallows (*Tachycineta bicolor*) (St. Louis and Breebaart 1991). Barrentine (1980) found that 36% of all grit particles retrieved from the stomachs of Barn Swallow (*Hirundo rustica*) nestlings between 1 and 16 days of age were calcareous. Parents appear to select and

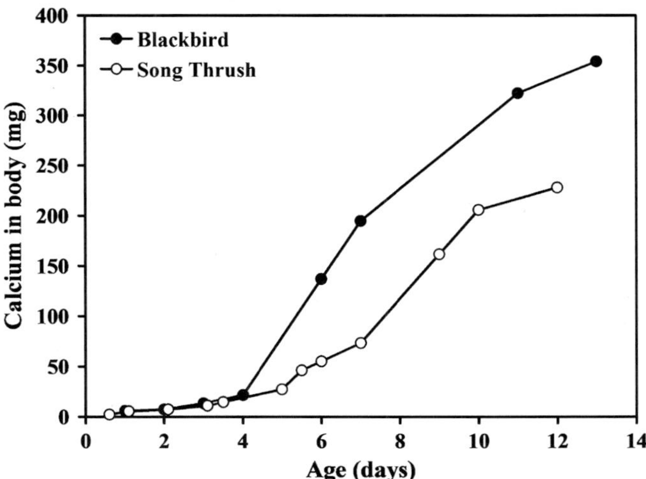

Fig. 2.1 Total mass of calcium in the bodies of Common Blackbirds and Song Thrushes sacrificed at various ages over the nestling period. Redrawn from Bilby and Widdowson (1971) by kind permission of "The British Journal of Nutrition"

offer these particles to their young before they reach 4 days of age. Such material may provide calcium for skeletal mineralization of nestlings, as well as aiding young birds in the mechanical digestion of insects in the gizzard. Grit and fragments of snail shell are delivered regularly to broods of Great, Blue, Coal (*Periparus ater*), and Willow (*Poecile montana*) Tits (Gibb and Betts 1963). In particular, these authors observed Coal Tits frequently adding grit to food bundles before delivering them to the nest. In the Great Tit, snail shell and grit were brought to the nest twice daily, each beakful of material being shared between approximately half the brood members (Betts 1955). The Eurasian Wryneck (*Jynx torquilla*) delivers many calcareous items to chicks, including stones, small bones and bone splinters, snail shell, and even pieces of china (Löhrl 1978; Klaver 1964). Furthermore, Löhrl (1978) reported that adult Eurasian Wrynecks do not carry off eggshells after hatching, feeding them instead to the young. This may be an adaptation to their particularly calcium-poor diet of ants and other insects fed to chicks.

Calcium-rich foods are fed to chicks soon after hatching. The young of Goldcrest (*Regulus regulus*) and Firecrest (*R. ignicapilla*) are fed with numerous pieces of snail shell and eggshell between 5 and 10 days of age (Thaler 1979). Betts (1955) found grit and snail shell in the diet of young tits for the entire 20-day nestling period, while Hågvar and Østbye (1976) found small bone fragments in stomachs of 10-day-old Meadow Pipit (*Anthus pratensis*) nestlings.

2.2.3   *Uptake and Utilization of Dietary Calcium During Egg Formation in the Domestic Chicken*

Most information concerning the calcium requirements for egg formation in birds concerns the Domestic Chicken, in which there is commercial interest in obtaining eggs with good quality shells. Hamilton (1982) suggested that eggshell problems resulted in an annual loss to the USA egg producer of around US $100 million.

Considerable research effort has been invested in supplementing foodstuffs with calcium-rich material and studying the effects of different feeding regimes on eggshell formation. The National Research Council (1984) recommended a daily intake of 3.75 g of calcium for laying domestic chickens if they are to produce eggs with good quality shells. An additional 1 g of calcium should be consumed per day by chickens that are old or by those with egg-quality problems that are particularly acute. However, providing diets of higher calcium content alone does not always improve eggshell quality (Keshavarz and Nakajima 1993). Indeed, supplying a surfeit of dietary calcium without maintaining a balance of other minerals can result in secondary deficiencies of other minerals important in avian nutrition and health (e.g., manganese, magnesium, and zinc; see Simons 1984). At high dietary levels of calcium (up to ~4%), calcium absorption is low (Hurwitz and Bar 1969), as is absorption of other minerals such as phosphorus (Hurwitz and Bar 1965).

Although it is important to supply poultry species with supplements of a sufficiently high calcium content, the amount of calcium that the animal can assimilate (i.e., the calcium bioavailability) should also be considered. The bioavailability of a nutrient, such as calcium, describes the extent of absorption and utilization of the nutrient in the food source of interest (Ammerman et al. 1995). Bioavailability is calculated by comparison with a standard reference material, with a bioavailability of 100% and in the same chemical form as the food source under test. For instance, the calcium concentration of dolomitic limestone is high, but calcium bioavailability is low because toxic levels of magnesium disrupt calcium uptake across the gut wall (Stillmak and Sunde 1971). The standard reference material used in the poultry industry is $CaCO_3$ (see Table 1 in Soares 1995). The bioavailability of some calcium supplements, such as bone meal, is 100% (Reid and Weber 1976), whereas steel slag and others is only 50% (Leach 1985). The poultry literature abounds with studies estimating the calcium bioavailability of poultry diets (e.g., Soares 1995) because both bone strength and eggshell thickness are easily measured in poultry species.

Crushed limestone and oyster shell are the most common calcium supplements added to poultry diets; the relative calcium bioavailabilities of both are 100%. Roland (1986) reviewed 44 papers that compared calcium availability of these two supplements, concluding that inclusion of large particles of either resulted in an improvement in eggshell quality. Scott et al. (1971) found that dietary substitution of 66% of crushed limestone for oyster shell resulted in a significant improvement in eggshell strength. This may be explained by a longer retention time of larger limestone particles in the gizzard than of smaller particles of oyster shell (Zhang and Coon 1997). During the night, when no further calcium is ingested, but when calcification of the eggshell occurs (Taylor 1970), calcium retained within the gizzard is "metered out" (Scott and Mullenhoff 1970) and chickens thereby do not become calcium deficient. Roland (1986) concluded that, regardless of the source of the calcium fed to laying chickens, supplying 33–66% of the larger particle size of $CaCO_3$, along with finely ground material, should result in eggs with good quality shells.

Domestic Chickens show a specific calcium appetite (Tordoff 2001; Joshua and Mueller 1979) that probably plays a fundamental role in controlling calcium levels during reproduction. Laying chickens appear to regulate food and calcium intake independently with suppression of the calcium appetite resulting from increased concentration of blood calcium. Consequently, chickens do not consume calcium at a constant rate throughout the day (Etches 1987) but calcium intake is tightly controlled during egg laying. Laying birds show a circadian rhythm of calcium intake (Tordoff 2001; Hughes 1972) that is related to their calcium requirements for eggshell formation (Burmester 1940). Sauveur and Mongin (1974) found oyster shell ingestion was greatest during the late afternoon, which ensured that dietary sources of calcium were available in the evening, when eggshell formation started. As the evening progressed and as dietary sources of calcium were exhausted, medullary bone reserves then provided calcium for continuing eggshell formation (Bloom et al. 1958). Work by Tyler (1954) confirmed this in laying chickens that were fed with radioactive calcium (^{45}Ca). Eggshells contained alternating layers of labeled and unlabeled calcium; these corresponded to calcium supplied by the diet in the afternoon, by the skeleton during the night, and by the diet once more during shell completion early in the morning.

Domestic Chickens lay eggs at progressively later hours on successive days of a laying sequence; the ovulatory cycle during laying can vary in length from just over 24–28.25 h (Etches and Schoch 1984). Most calcium deposition into eggshells occurs in the 15-h period preceding laying. Deposition of total eggshell calcium is 8.9%/h for the first 4 h of the period and 5.6%/h thereafter (Talbot and Tyler 1974). The fully formed egg of the Domestic Chicken contains ~5 g of $CaCO_3$. One of the most remarkable features of egg formation in Domestic Chickens is the rate of calcium deposition (i.e. ~125 mg/h). However, at any instant, there are only 25 mg of calcium in circulation (Taylor 1970) so blood calcium is entirely cleared once every 12 min. The Domestic Chicken is an extreme example of an avian species that is capable of mobilizing up to 10% of total body calcium in a 24-h period during egg laying (Taylor 1965). In the Domestic Chicken, the annual calcium turnover as eggshells is 30 times the bird's total somatic calcium pool at any given time (Gilbert 1983).

2.2.4 Uptake and Utilization of Dietary Calcium During Egg Formation in Other Avian Species

Resident passerines, such as Great Tits, are entirely dependent upon calcium-rich foods for the calcium required for egg formation. Skeletons of birds generally do not maintain long-term stores of calcium for the production of eggs (e.g., Graveland and van Gijzen 1994). Some birds, such as Neotropical migrants that migrate over long distances and over-winter in calcium-rich ecosystems on $CaCO_3$ bedrock, might be expected to deposit calcium reserves for egg formation before migration. However, Blum et al. (2001) showed that Black-throated Blue Warblers (*Dendroica caerulescens*) relied wholly upon calcium ingested on breeding territories. They examined the strontium (Sr) signatures of bedrock from breeding (northeastern USA) and over-wintering (Caribbean islands) ranges of these birds. Isotopic signatures of skeletons of adult birds, eggshells, egg contents, and food sources were also examined by using strontium as a surrogate for calcium because of its similar pathways of uptake and utilization. Comparisons of $^{87}Sr/^{86}Sr$ ratios of food sources on breeding territories with $^{87}Sr/^{86}Sr$ ratios of egg constituents indicated that egg products were obtained exclusively from sources in the breeding areas and not from calcium-rich sources in the wintering grounds. Furthermore, Blum et al. (2001) corroborated the results reported by Pahl et al. (1997), finding that breeding females contributed little calcium from their skeletons to egg contents or shells.

It is not surprising that migrants, like the Black-throated Blue Warbler, do not deposit calcium as skeletal reserves before migration. Wintering females of this species have an average body mass of 8.7 g (Holmes 1994) and from the allometric equation of Graveland (1995)

$$\text{Skeletal mass } (g) = 0.0339 \times (\text{body mass})^{1.07}, \qquad (2.1)$$

we calculate that the average skeletal mass of the female is 0.343 g. An average eggshell contains ~20 mg of calcium (Taliaferro et al. 2001). A clutch of four eggs (i.e., modal clutch size; Holmes 1994) will therefore contain ~80 mg of calcium. Since medullary bone (a highly labile store of calcium) does not form in the long bones of reproductive females until about 10 days before egg laying (Johnson 2000), calcium destined for the eggshell would have to be deposited as cortical (structural) bone, which is comprised of 22.5% calcium (Graveland 1995). The female would therefore carry an extra 356 mg of stored calcium in cortical bone from wintering to breeding grounds. This represents an increase in wintering body mass of ~4%. Although Black-throated Blue Warblers refuel at several stopovers during migration, such an increase in body mass would nevertheless have major implications for flight performance. Whereas the fuel load (i.e., fat) is accumulated before departure but lost progressively during migration (e.g., Bairlein and Simons 1995), extra-skeletal reserves of calcium would remain constant throughout migration. The resultant changes in overall body mass would affect wingbeat kinematics (e.g., Videler 1995) and ultimately the flight ranges of migrants (Biebach 1992). This would affect the entire temporal and spatial organization of migration (Newton 2008; Schaub and Jenni 2000).

Of the species studied to date, most appear incapable of maintaining long-term skeletal calcium reserves in readiness for egg production (but see Reynolds 2003; Larison et al. 2001). Consequently, they increase their dietary intake of calcium a few days prior to the onset of laying. However, there are a few exceptions. The Red Knot (*Calidris canutus*) stores significant amounts of calcium in its skeleton before laying (Piersma et al. 1996). This storage starts a few weeks before laying; skeletal mass increases by 30–50% during this period, allowing birds to produce two of their four-egg clutch without ingesting any calcium during laying. Further depletion of medullary bone reserves, similar to the extent found in other birds (Graveland 1995), provides sufficient calcium for the formation of another eggshell.

Depending upon the timing of breeding, the Ruddy Duck (*Oxyura jamaicensis*) is also capable of mobilizing calcium from endogenous reserves for egg formation. Alisauskas and Ankney (1994) found that mineral reserves declined progressively with eggshell formation in early breeding birds, which lost 0.08 g of mineral reserves for every gram of eggshell deposited. This was not the case for later-breeding birds, suggesting that the later breeders relied more heavily on exogenous calcium sources. Even in the birds that breed later in the season, endogenous reserves of calcium may assist the laying bird in buffering temporary shortfalls in dietary calcium. Alisauskas and Ankney (1994) also showed that endogenous calcium reserves of the female Ruddy Duck might influence clutch size; the number of developing follicles was correlated with mineral reserves. In other waterfowl species, the number of developing follicles is correlated with fat (e.g., Northern Shoveler [*Anas clypeata*], Ankney and Afton 1988) and protein reserves (e.g., Gadwall [*A. strepera*], Ankney and Alisauskas 1991). Compared with other species, the eggshell of the Ruddy Duck represents a significantly higher proportion of somatic mineral reserves of the female. Nutrient-reserve thresholds for the onset of egg laying have been demonstrated for the Ring-necked Duck (Alisauskas et al. 1990) and the Gadwall (Ankney and Alisauskas 1991). However, in the Ruddy Duck, the storage of not only fat and protein, but also calcium, during the prelaying period has allowed this species to produce large eggs at a rapid rate, sometimes even when calcium-rich food is scarce. In time, as the nutritional strategies of breeding birds are investigated further, we may discover that other species also supply calcium for egg formation by relying heavily upon skeletal reserves that are deposited some time before egg laying begins. For example, many Arctic nesting geese start to lay within a few days of arrival on the breeding grounds (Raveling 1978), well before they can accumulate calcium by intensive foraging. Therefore, Ryder (1970) and Ankney and MacInnes (1978) proposed that clutch size might be determined by the size of the nutrient reserves of females arriving on the breeding grounds.

Extensive information is available on the calcium logistics of egg production in poultry species because calcium intake of laying birds can be closely monitored. Similar quantitative data are difficult to collect for free-living birds, but it is possible to make crude calculations of dietary calcium intake of captive wild birds during egg laying. Graveland and Berends (1997) found that daily snail shell consumption by captive Great Tits was minimal (4 mg) but constant in the 2-week prelaying period (Fig. 2.2). Daily consumption then increased to 16 mg 2 days

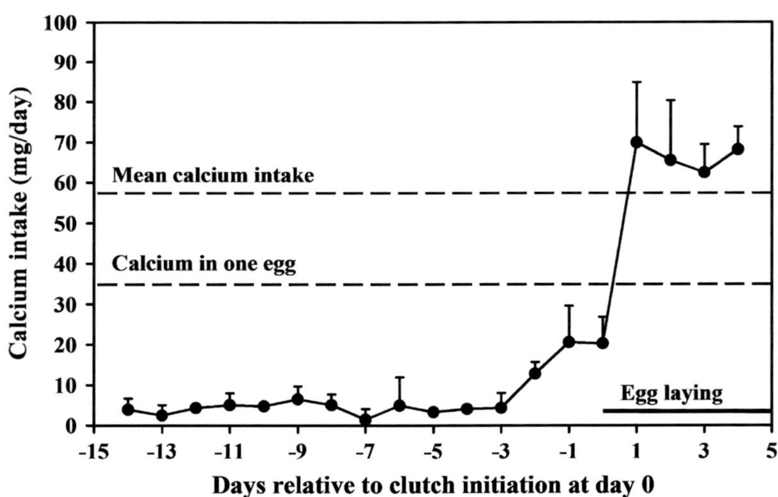

Fig. 2.2 Calcium consumption (in the form of snail shells) (+ 1 SE) by female Great Tits in relation to the onset of egg laying. Calcium intake is calculated from intake of shell fragments containing 33.1% calcium. Daily consumption is measured over a 24-h period beginning at 1300 h on the day before an egg was laid. Redrawn from Graveland and Berends (1997) by kind permission of "Physiological Zoology"

before the first egg was laid and then increased to a daily average of ~65 mg during egg laying. Captive Zebra Finches also show a pronounced daily increase in cuttlefish bone consumption from 50 mg (15 mg calcium) to 185 mg (54 mg calcium) 2 days before clutch initiation (Houston et al. 1995). The skeleton of a female Zebra Finch contains 90 mg of calcium, and a four-egg clutch contains 71 mg of calcium (Houston et al. 1995). There is little skeletal contribution of calcium for eggshell formation; an average Zebra Finch egg contains 17.8 mg of calcium, and birds derive over 20 mg of calcium per day from ingested cuttlefish bone in the laboratory (Houston et al. 1995).

House Sparrows (*Passer domesticus*) do not store calcium during rapid follicular growth, nor do they deplete endogenous calcium reserves during clutch formation (Pinowska and Kraśnicki 1985b). Krementz and Ankney (1995) found that clutch size was independent of endogenous reserves of calcium (see also Anderson 1995), despite an increase in total body calcium before egg production (Schifferli 1979) levels declined after clutch completion. They proposed that laying birds deposit ingested calcium as medullary bone and utilize it for overnight calcification of the egg in the same way as Domestic Chickens (Simkiss 1967), that is, maintain a short-term, labile store of calcium that is depleted and replenished on a daily basis during laying. This view is supported by the foraging patterns of laying House Sparrows in which fragments of snail shells were found more often in the gizzards of birds collected in the late afternoon and early evening than in those collected before midday (Schifferli 1976). Many species that lay early in the morning show an evening peak in consumption of calcium-rich foods (Graveland and Berends 1997).

Reynolds (1997) provided further evidence that small passerines utilize dietary calcium during laying in a way similar to that of Domestic Chickens. As in the dosing experiments carried out by poultry scientists (Clunies et al. 1993), Reynolds (1997) administered ^{45}Ca to Zebra Finch females exactly 1 h after laying. Laying birds excreted significantly less calcium across the entire ovulatory cycle than did nonlaying (control) birds that were dosed at a similar time (Fig. 2.3a). Furthermore, in the first 16 h postdosing, laying birds deposited significantly more calcium in their skeletons than controls, but deposition of dietary calcium was similar for nonlaying and laying birds in the 7-h period preceding the next oviposition by the laying birds (Fig. 2.3b). Dietary calcium was 60 times more abundant in the reproductive tissues of laying birds than in controls 8–16 h following dosing, a time when calcification of the eggshell was occurring (Fig. 2.3c).

Microscopic examination of bone structure of laying birds reveals the marked changes in bone cell types and bone tissues that occur during a single laying cycle. Reynolds (1998) compared longitudinal 6-μm sections from long bones of laying Zebra Finches with those from nonlaying (control) birds. Bone composition of laying birds changed profoundly with time postlaying, and changes in relative bone cell numbers and degree of calcification could be related to the changes in calcium demand during egg formation. Osteoid tissue, the collagenous matrix laid down by osteoblasts prior to calcification, was abundant in sections from laying hens. Number of osteoblast (osteogenic) and osteoclast (osteolytic) cells were significantly higher in laying than in control sections, indicating that bone from laying

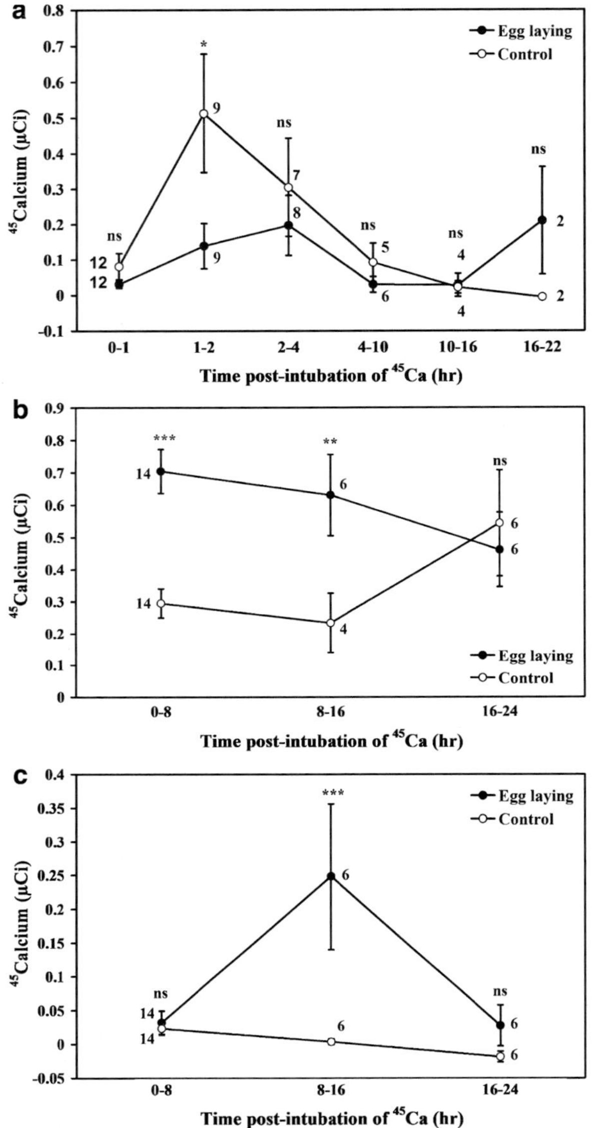

Fig. 2.3 ^{45}Ca content (± 1 SE) in (**a**) feces, (**b**) skeletons, and (**c**) ovarian tissues of pairs of egg laying and control Zebra Finches dosed with ^{45}Ca 1 h after oviposition in egg layers. Numbers represent sample sizes. Comparisons within time periods between egg layers and controls: ns, $P>0.05$; *, $P<0.025$; **, $P<0.01$; ***, $P<0.0005$. Reproduced from Reynolds (1997) by kind permission of "The Auk"

females was metabolically active. In contrast, all sections from control hens were heavily calcified, possessed little osteoid tissue, were metabolically inactive, and had low numbers of osteoblasts and osteoclasts.

2.2.5 Calcium and Eggshell Pigmentation

Avian eggshells display a broad diversity of coloration, both in terms of background base color and the patterns of pigmentation or maculation, and a number of hypotheses have been proposed to explain its functional significance (reviewed in Kilner 2006). Despite over a century of discussion on the subject, a structural role for such coloration has only recently been proposed. Among a number of structural explanations of eggshell coloration (reviewed in Reynolds et al. 2009), one is directly related to dietary calcium. Gosler et al. (2000) established that eggshell pigmentation patterns in a population of Great Tits were subject to female sex-linked inheritance but their functional significance remained obscure. In a follow-up study of the same population, Gosler et al. (2005) found that pigmentation patterns on eggshells were localized where eggshell was thin and that they were negatively correlated with soil calcium availability near the nest. Thus, they proposed a compensatory role for eggshell maculation (and, specifically, for protoporphyrin pigments) that might reinforce the thin eggshell that is formed by birds breeding in areas of low calcium availability. Solomon (1997) was the first to propose such a structural reinforcement function for porphyrins in Domestic Chicken eggshells because of their shared structural similarities to phthalocyanines that are lubricants in solid state engineering. Much remains to be studied in this area before such a structural function of eggshell coloration receives widespread acceptance but the initial findings relating calcium availability to pigmentation patterns on eggshells of Great Tits are intriguing and will surely stimulate extensive future investigations.

2.3 Instances of Calcium-Limited Reproduction in Birds

Drent and Woldendorp (1989) were the first to investigate the effects of acid rain on avian reproduction in forest ecosystems. They found a dramatic rise in the incidence of eggshell defects in clutches of Great Tits over a 6-year period (1983–1988), especially in areas of poor soil quality. Eggshell defects were also identified in clutches of Blue and Coal Tits, Wood Nuthatches (*Sitta europaea*), and Great Spotted Woodpeckers (*Dendrocopos major*). However, there were no egg defects in clutches of European Pied Flycatchers (*Ficedula hypoleuca*) that fed on calcium-rich foods (e.g., millipedes [*Diplopoda* spp.] and woodlice [*Isopoda* spp.]) in addition to their normal insect prey throughout the summer (Dekhuijzen and Schuijl 1996). These arthropods contain 100 times more calcium than most arthropods in the flycatchers' diet (Graveland and van Gijzen 1994). Therefore, compared with Great Tits, which obtain only 5–10% of calcium for clutch formation from seeds and arthropods, European Pied Flycatchers do not have to change their diet significantly during breeding to satisfy the calcium demands of egg laying.

Graveland et al. (1994) investigated the effects of calcium deficiency on the reproductive performance of Great Tits in The Netherlands. They explained the increased incidence of eggshell defects, nest desertion, and empty nests in terms of poor availability of snail shells, the main source of calcium for laying Great Tits. The effects of reduced calcium availability on the reproductive output of birds were particularly pronounced in areas of poor soil quality, where snails declined as acid rain leached calcium from soils. Eggshell quality was ameliorated in areas of nutrient-poor soils when birds were artificially provided with calcium (e.g., grit, eggshells) (Graveland and Drent 1997; Graveland 1996a).

Since the studies of Drent and Woldendorp (1989) and Graveland et al. (1994), calcium-limited reproduction has been reported in a number of other species (Table 2.2), particularly in those nesting in acidified areas. Blancher and McAuley (1987) reviewed the evidence for reduced breeding success in North American

Table 2.2 Evidence of calcium-limited reproduction in birds

Species	Evidence	Reference
Black Tern (*Chlidonias niger*)	Incomplete clutches, hatching failure, thin-shelled eggs	Beintema et al. 1997
Great Spotted Woodpecker (*Dendrocopos major*)	Thin-shelled eggs	Drent and Woldendorp 1989
Tree Swallow (*Tachycineta bicolor*)	Reduced egg size and clutch volume	Blancher and McNicol 1988
	Reduced egg size and hatching success (nestlings fledged/ nest box)	St. Louis and Barlow 1993
Great Tit (*Parus major*)	Incubating empty nests	Zang 1998; Schmidt and Zitzmann 1990; Winkel and Hudde 1990;
	Reduced clutch size and recruitment, abnormal shell structure causing dehydration	Wilkin et al. 2009; Weimer and Schmidt 1998; Carlsson et al. 1991
	Shell defects, increased clutch desertion	Pinxten and Eens 1997; Graveland et al. 1994
Blue Tit (*Cyanistes caeruleus*)	Reduced breeding success, reduced number of fledglings	Dekhuijzen and Schuijl 1996
	Incubating empty nests	Schmidt and Zitzmann 1990
Wood Nuthatch (*Sitta europaea*)	Thin-shelled eggs	Drent and Woldendorp 1989
White-throated Dipper (*Cinclus cinclus*)	Thin-shelled eggs	Ormerod et al. 1991
Meadow Pipit (*Anthus pratensis*)	Thin-shelled eggs	Bureš and Weidinger 2001
European Pied (*Ficedula hypoleuca*) and Collared (*F. albicollis*) Flycatcher	Irregular laying, reduced clutch size, shell defects, desiccated eggs during incubation	Bureš and Weidinger 2003
Eastern Kingbird (*Tyrannus tyrannus*)	Retarded laying date; reduced brood and clutch size. Reduced hatching success, eggs with thin permeable shells	Glooschenko et al. 1986

birds nesting in acidic wetlands. All studies demonstrated detrimental effects of acidification on food availability. Aquatic invertebrates lost salts and showed poor calcium uptake in acidic conditions (Malley 1980), and those with calcareous shells (e.g., amphipods, clams, snails) were absent in waters of low pH (Mills and Schindler 1986). In Tree Swallows, a number of reproductive parameters were positively related to wetland pH: clutch size, egg volume, and fledging success (Blancher and McNicol 1988); egg size and nestling growth (St. Louis and Barlow 1993); and the availability of calcium-rich foods (Blancher and McNicol 1991). Similar results were found in Eastern Kingbirds (*Tyrannus tyrannus*), in which individuals nesting near acidic wetlands produced thinner shelled eggs and nestlings with poorer growth rates than individuals nesting in buffered areas (Glooschenko et al. 1986). White-throated Dippers (*Cinclus cinclus*) have been particularly vulnerable to increasing acidity of stream habitats. In comparisons of White-throated Dippers nesting along acidic streams and along streams close to neutral, Ormerod et al. (1991) found that the former delayed breeding, reduced clutch and brood size, and suffered retarded chick growth. Egg mass and eggshell thickness also declined at more acidic sites (Ormerod et al. 1988). The scarcity of calcium-rich foods (e.g., fish, mollusks, shrimps) was directly responsible for declines in the breeding success of White-throated Dippers along acidic streams.

Calcium-limited reproduction is rare, but not absent, in species that routinely consume calcium-rich foods during the nonbreeding season. Some piscivores can be calcium-stressed during laying. Female Sandwich Terns (*Sterna sandvicensis*) require 1.4 g of calcium for the formation of their two-egg clutches (Brenninkmeijer et al. 1997). They obtain a maximum of 0.56 g of calcium from skeletal reserves but only 0.6 g from the regular diet of teleost fish. Brenninkmeijer et al. (1997) observed at least five birds eating shell fragments at the onset of egg laying, presumably to make up for the shortfall of 0.24 g still required for clutch formation. Likewise, Nisbet (1997) reported shell eating in female Common Terns (*S. hirundo*), but only in the 3–5-day period before the onset of egg laying and the day of final oviposition. Shell fragments probably restored the calcium balance during egg laying at a time when the birds were calcium-stressed as a result of high calcium demand.

2.4 Effects of Limited Calcium Availability

Calcium deficiency can result in a whole suite of symptoms in birds (Robbins 1993 and references therein). Severe deficiency occurs most commonly in captive species (e.g., Wallach and Flieg 1969) that have been selected for high egg production, but it has been documented in free-living species too (e.g., Graveland 1996b). Generally, wild birds are less abundant in areas where calcium availability is insufficient to meet the requirements for self-maintenance and reproduction (e.g., Common Pheasant; Dale 1954); where wild birds do breed successfully, clutch size may be constrained by broad-scale variation in dietary calcium as suggested by Patten

(2007). Compared with captive birds, detection of the more subtle effects of calcium deficiency remains problematic in free-ranging individuals.

Mobilization of calcium during reproduction is mediated through the endocrine system (Scanes et al. 1982). Reproductive females show a hypercalcemic response when parathyroid hormone (PTH) brings about an increase in plasma calcium concentration by increasing calcium resorption from renal tubules and medullary bone (Dacke et al. 1993). PTH also acts on the kidneys to stimulate the release of the hormonal version of vitamin D_3, 1,25-dihydrocholecalciferol or $1,25(OH)_2D_3$, which, in turn, increases calcium absorption from the gut and calcium resorption from medullary bone (Soares 1984). The hormone calcitonin (CT) is secreted from the ultimobranchial glands of birds in response to elevated plasma calcium concentrations that occur during egg formation. The role of CT in calcium mobilization during egg laying remains ill-defined (Hirsch and Baruch 2003), but it appears that CT controls the level of hypercalcemia protecting the skeleton from excessive calcium resorption (Baimbridge and Taylor 1981).

In normocalcemic birds, plasma calcium concentrations increase dramatically a few days prior to ovulation (e.g., from 2.2 mmol/L to 5.0 mmol/L at ovulation in the Common Pigeon [*Columba livia*]; Lumeij 1994). These reflect increases in the concentration of protein-bound calcium and in calcium that is complexed with yolk precursor proteins that are released by the liver in response to estrogen. Egg production is therefore carefully controlled according to calcium availability from dietary and endogenous sources. However, when calcium is limiting, maintenance of skeletal integrity takes precedence over the requirements for reproduction. Luck and Scanes (1982) showed that restricted access to calcium caused Domestic Chickens to cease laying before the structural integrity of cortical bone was compromised. Similar restrictions in calcium availability to laying wild birds can result in a number of physical abnormalities in developing eggs and young, ultimately affecting reproductive success, as reported in Table 2.2. As in Domestic Chickens, a bird that fails to find sufficient calcium for egg formation on its breeding territory may not lay (e.g., Carlsson et al. 1991), but more often calcium-stressed birds lay eggs with abnormal shells (Graveland 1995 and references therein). Subsequently, the most obvious result of calcium limitation is reproductive failure caused by desertion of nests and failure to incubate defective eggs. For example, Graveland (1996a) reported that 43% of female Great Tits producing abnormal eggs deserted, usually during laying, compared with 16% of birds laying normal eggs.

2.4.1 Adult Physiology

Nutritional calcium deficiency can disrupt skeletal integrity in laying birds by causing osteomalacia, also called osteodystrophy. Parathyroid glands become enlarged and secrete excess PTH, which activates osteoclasts, the osteolytic bone cells that are responsible for calcium resorption. The result is a total demineralization of the medullary, and then the cortical, bone, the latter becoming so thin that "greenstick fractures" commonly occur. Grey Parrots (*Psittacus erithacus*) respond to nutritional calcium deficiency by developing hypocalcemia syndrome (Lumeij 1994), but, strangely, this does not result in skeletal demineralization. Serum calcium levels decrease and birds become uncoordinated, convulsing and frequently falling from perches. Their inability to mobilize skeletal calcium reserves may be caused by a viral disruption of parathyroid function. The higher frequency of hypocalcemia in African parrots, as compared with Amazon parrots, may result from the former having significantly lower plasma calcium concentrations (Lumeij 1990). Interestingly, evidence is starting to emerge that geophagy is necessary for African parrots if they are to obtain sufficient essential minerals, such as calcium (May 2001); in contrast, South American parrots consume soil of only moderate calcium content (but see Foster 2007). The high cation exchange capacity and high content of cation-binding minerals of the soils consumed by the latter suggest that geophagy is employed by these parrots only to bind toxins and secondary plant compounds (see Diamond et al. 1999; Gilardi et al. 1999), and not to enhance their breeding performance per se.

Females of all avian species are physiologically osteoporotic as a result of depletion of medullary bone reserves in order to provide calcium for the formation of the eggshell. Osteoporosis, also known as cage layer fatigue, is the most important of the skeletal diseases in Domestic Chickens, particularly in those that sustain high egg production. Osteoporosis, resulting in skeletal abnormalities and paralysis, is most obvious in birds that are restricted in their movement by confinement in small enclosures and those that receive a calcium-deficient diet.

Nutritional deficiency of calcium can also result in the abnormal passage of eggs through the reproductive tract, particularly in those species prone to bouts of chronic egg laying (e.g., domesticated psittacine birds). When an egg fails to pass through the oviduct at the normal rate, a condition called egg binding can develop. The partially calcified egg becomes lodged in the cloaca and cannot be laid without remedial action. Rapid displacement of the egg from the pelvic canal is imperative because egg binding can result in circulatory disorders and shock as the egg compresses pelvic blood vessels and the kidneys. Furthermore, the partially calcified egg can break inside the bird, resulting in the worst cases in death from peritonitis. The onset of egg binding can be rapid and the result fatal (Quesenberry and Hillyer 1994), but, except for isolated cases, little is known about the mortality of wild birds (e.g., Chattopadhyay 1980) as a result of this condition.

Both dysfunction of oviduct muscle (e.g., from calcium metabolic disease), and the laying of malformed eggs (e.g., from dietary calcium deficiency), can cause dystocia. In this condition, the developing egg either blocks the cloaca or it causes prolapse of the oviduct through the cloacal opening. Egg binding and dystocia are found most commonly in domesticated birds that are subjected to high egg production over a wide range of environmental conditions, which probably results in disruption of hormonal control over the laying process.

2.4.2   Eggshell Structure

In those eggs that are incubated despite shell abnormalities, embryonic death probably occurs primarily as a result of desiccation. Incomplete calcification resulted in enlarged pores in the shells of European Pied Flycatcher eggs (Nyholm and Myhrberg 1977). Substantial volumes of water can be lost during incubation, even in eggs with complete, but thin shells (Ar et al. 1974). Abnormal shell structure might also result in more subtle effects on the osmoregulation of embryos (e.g., Higham and Gosler 2006). Embryos mobilize calcium from the eggshell (Karlsson and Lilja 2008), and those encased by abnormal shells suffer hypocalcemia. As a result, they cannot regulate fluid volume because of ionic imbalance in the allantoic sac (Hoyt 1979), water loss at inappropriate stages of incubation (Ar and Rahn 1980), and kidney malfunction. In addition, embryonic growth is retarded in eggs with abnormal shells (Dunn and Boone 1977).

2.4.3   Chick Development

In addition to causing production of aberrant eggshells, limited calcium availability can also affect the growth of chicks. There is evidence that in primarily insectivorous Black Terns (*Chlidonias niger*), females may be calcium-stressed during egg production and that limited calcium availability may also affect chick growth. Although not widely reported in the literature, males have been seen feeding females with fish during courtship (Oring, unpublished observations) in Idaho, USA, perhaps providing them with additional dietary calcium in preparation for egg laying. Beintema et al. (1997) reported that Black Terns nesting on acid bogs in The Netherlands were feeding chicks exclusively on aquatic larvae and insects, both of which were low in calcium. All chicks in one nest were suffering from severe rickets, and all died by 25 days of age. Calcium limitation was confirmed as the principal cause of skeletal under-development when Black Tern chicks were force-fed calcium pills three times a week throughout the fledging period. Calcium-supplemented chicks were ~16 g heavier than controls (unsupplemented), the latter developing rachitic symptoms in wing and leg bones. In the best-documented case of calcium deficiency in growing nestlings, Richardson et al. (1986) found that in 1977 up to 16.9% of nestling Cape Vultures at Magaliesberg and Botswana colonies exhibited metabolic bone disease or rickets. Their skeletons were poorly calcified (excessive flexibility in wing bones resulted in numerous fractures) because the parents failed to find sufficient bone fragments of a size small enough for young birds to swallow. A similar situation may have resulted in the death of a Red Kite (*Milvus milvus*) chick in The Midlands, UK; it also exhibited rachitic symptoms (Fig. 2.4).

Even apparently normal eggs that appear to show no aberrant eggshell structure may harbor impoverished developmental conditions for chicks that may be linked

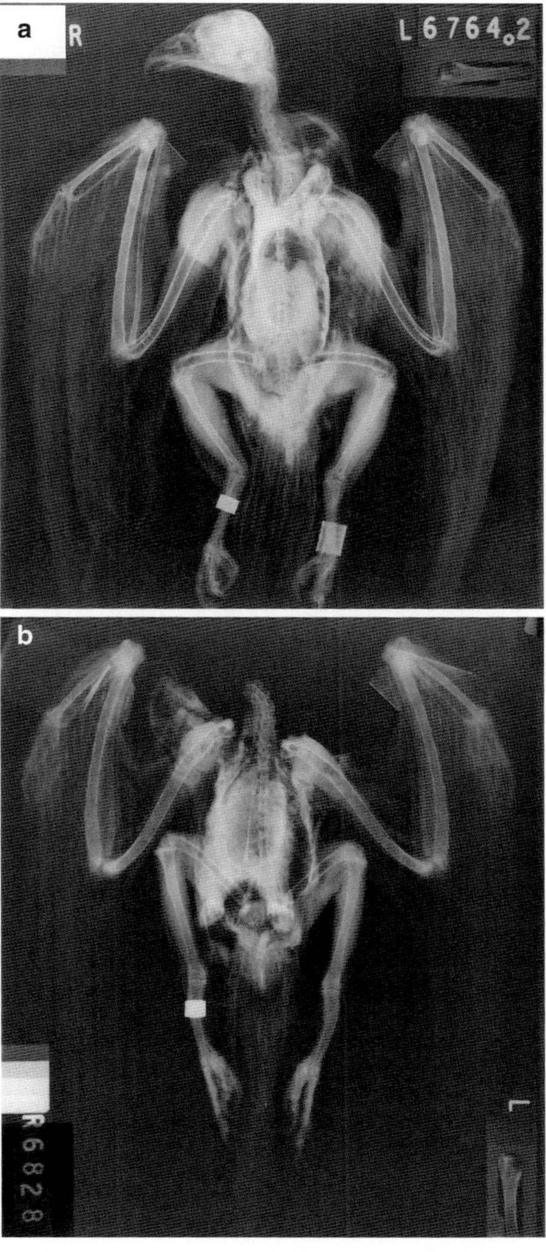

Fig. 2.4 X-rays of Red Kite chicks showing (**a**) normal skeletal development resulting from a calcium-rich diet and (**b**) abnormal skeletal mineralization resulting from a calcium-poor diet. In the latter, note the curved (deformed) tibiotarsi and the thin cortices of the leg bones in the 9-week-old bird that was found dead near the nest. This bird died from metabolic bone disease resulting from an inability to consume sufficient calcium-rich material (usually small bone fragments). Reproduced by kind permission of Natural England and the Institute of Zoology at the Zoological Society of London

to poor calcium supply from egg constituents other than the shell. Tilgar et al. (2005) supplemented Great Tits breeding in base-poor Estonian forests with calcium and showed that egg yolk calcium concentration was elevated and that this was positively related to growth of nestling tarsus length but only in the first half of the nestling period. Such calcium limitation might produce particularly acute depression of chick growth in years when food availability is low, while remaining difficult to detect through "traditional measures" of reproductive calcium limitation (see Table 2.2).

2.4.4 Adult Behavior

In breeding adults, limited calcium availability can produce subtle changes in behavior. With diminishing calcium availability, foraging birds become more selective, favoring calcium-rich items. For example, a reduction of dietary calcium from 2.65 to 0.50% in laying Common Pheasants led to a 400% increase in grit consumption (Sadler 1961); the birds favored calcium-rich limestone over granite grit. With progressive calcium deficiency, some birds increasingly attempt to consume bizarre objects in response to a calcium-specific appetite (Tordoff 2001). In ranched areas of South Africa where large carnivores were scarce, Griffon Vultures (*Gyps fulvus*) obtained few bone fragments small enough to feed to their chicks, and Mundy and Ledger (1976) found fragments of china, glass, metal, and plastic in their nests. Depriving captive Great Tits of snail shells during egg laying resulted in a doubling of search effort for dietary calcium as compared with tits that had access to fragments of snail shell (Graveland and Berends 1997). Deprivation also resulted in alterations in other behavior patterns, with females searching for calcium at the expense of maintenance behaviors (i.e., perching, preening, and sleeping) and consuming more clay, sand, and small stones than birds provided with snail shell. Graveland and Berends (1997) found also that the majority (81.8%) of the calcium-deprived females consumed their own eggs. Strangely, two of these calcium-deprived females also built second nests and proceeded to lay eggs in both. Work on calcium-specific foraging behavior of Great Tits has recently been extended to free-living birds where an analysis of 41 years of breeding data at Wytham Woods, Oxford, UK, revealed that females may forage interterritorially (i.e., >300 m from the nest) during egg laying to obtain sufficient calcium for egg formation (Wilkin et al. 2009).

2.5 Detection of a Dietary Calcium Threshold

Table 2.2 lists a number of studies that have found evidence that calcium limits reproduction in birds. In addition many species may have to sacrifice valuable foraging time for other nutrients to forage solely for calcium. Grey-backed Camaropteras

(*Camaroptera brevicaudata*) lost fat reserves during laying as a result of investment in time spent foraging for calcium (Fogden and Fogden 1979). Egg laying Sand Martins (*Riparia riparia*) and Barn Swallows needed to forage for <6 h per day to meet their requirements for energy, protein, and sulfur-amino acids, but they found it difficult to collect sufficient calcium to meet the demands of daily maintenance and egg production (Turner 1982).

Identifying dietary calcium thresholds for free-living birds may prove difficult. Ramsay and Houston (1999) found that Blue Tits nesting near Loch Lomond, in west-central Scotland, were able to find sufficient calcium for clutch formation, although they nested in an area where exchangeable soil calcium (0.02 mg/g^2) and snail abundance (0.36 snails/m^2) were significantly lower than in the Buunderkamp forest in The Netherlands, where birds were severely calcium-limited (Graveland 1995). A search of 22 m^2 of leaf litter and topsoil from the Scottish study area found only eight snail shells (Ramsay and Houston 1999). However, even in habitats where calcium availability was low, breeding birds obtained sufficient calcium for egg production. Their gizzards contained calcium-rich material (Ramsay and Houston 1999) and they laid eggs without defective shells. These results suggest that a major limitation currently is our inability to sample calcium availability accurately. Ramsay and Houston (1999) sampled leaf litter and topsoil but birds may have been taking snails from other sites. The main source of calcium for breeding Black-throated Blue Warblers in New England, USA, forests is land snails that are only ~1–2 mm in length (Taliaferro et al. 2001). These are present in the leaf litter, in decaying wood, and under logs making them virtually impossible to detect by researchers.

Studying captive birds in order to identify dietary calcium thresholds may be easier than field studies because calcium availability to captives can be controlled and foraging efforts can be quantified. For example, Reynolds (2001) investigated the effects of low dietary calcium on egg and laying parameters of captive Zebra Finches. Birds on low calcium did not reduce clutch size below that of birds on ad libitum calcium, but deposition of calcium in eggshells, measured by shell ash mass, declined with laying order (Fig. 2.5). However, captive birds cannot forage over wide areas to redress calcium deficits that might arise during laying. Detection of dietary thresholds without causing birds to become egg bound may be achievable only through subtle, incremental reductions in calcium availability.

2.6 Causes of Reductions in Calcium Availability

2.6.1 Acidification

Many of the studies that have reported calcium-limited reproduction of wild birds have attributed their results to "acid rain," the primary cause of calcium loss from the environment. Although acidification may have caused the most obvious depletion

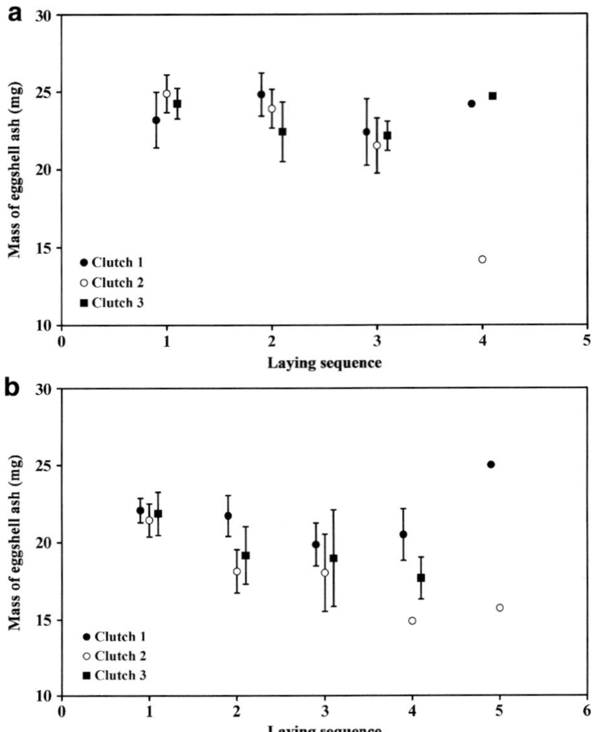

Fig. 2.5 Masses of eggshell ash (± 1 SE) through the egg-laying sequence for Zebra Finch females on (**a**) a low calcium diet for clutch 2 and on ad libitum calcium for clutches 1 and 3, and on (**b**) ad libitum calcium (control) for all three clutches. Reproduced from Reynolds (2001) by kind permission of "Ibis"

of environmental calcium (Graveland 1998 and references therein) in the past, there is increasing evidence that reducing emissions of nitrous and sulfurous oxides and ammonia from anthropogenic sources slows rates of acidification (Hildrew and Ormerod 1995). Stoddard et al. (1999) reported significant declines in the rates of acid deposition in the 1980s and 1990s across many regions of Europe and North America. As a result, acidified, aquatic ecosystems throughout north-central Europe have gradually recovered (i.e., increased alkalinity), especially in the 1990s, but this occurred in only one monitored region of North America. Whereas decreases in sulfate ions led to this recovery in Europe in the 1990s, in North America declines in sulfate ions were insufficient to counteract the continued leaching of basic cations, such as calcium, from sensitive soils. As a result, three extensive regions of North America (south/central Ontario, Adirondack/Catskill Mountains, and mid-western North America) remain acidified despite continued declines in acid deposition. Nevertheless, in areas of the world where the emissions of precursors of acid rain have declined, the adverse effects of acidification on terrestrial and aquatic ecosystems have been mitigated and amelioration persists to the present day (e.g., Singh and Agrawal 2008).

Mahony et al. (1997) compared breeding success of Black-capped Chickadees (*Poecile atricapillus*) and Chestnut-sided Warblers (*Dendroica pensylvanica*) at two forest sites, one healthy and the other declining (acidified) in south/central Ontario. Although they found no differences in clutch or brood sizes or hatching success for either species between sites, they reported chickadees foraging consistently lower in the canopy at the acidified site, perhaps in response to dieback and the consequent declines in prey abundance higher in the canopy. More studies of this kind are needed if the long-term effects of sustained acidic conditions on the breeding performance of birds are to be appreciated fully. Although Graveland (1998) reviewed the effects of acidification on avian reproduction, investigations of long-term effects of acidification on calcium availability and the subsequent breeding performance of avian species nesting over wide spatial scales are few in number (e.g., Hames et al. 2002; Vickery and Chamberlain 1999).

2.6.2 Agricultural Intensification

Agricultural intensification may have resulted in a decrease of calcium availability. The Netherlands suffers from the adverse environmental effects of both heavy industry and intensive agriculture, resulting in large tracts of land that are now calcium-impoverished (Graveland, personal communication). Historically, forest ecosystems have undergone nutrient depletion as livestock grazed in forests during the day but spent nights in farm buildings. The manure produced at night was subsequently applied to arable land and not recycled into forests. Dutch forests now support few calcium-rich organisms as food for breeding birds. Agricultural intensification in the UK too has resulted in the wholesale decline of the Corn Bunting (*Emberiza calandra*), its abundance being positively correlated with the proportion of arable land (Shrubb 1997). Soil type may also be crucial. Birds avoid clay areas of central East Anglia and southeast England. Reports from other areas of England suggest that birds are found only in clay areas when they are displaced from calcareous soils by conspecifics. Gillings and Watts (1997) found that Corn Buntings preferred light, free-draining soils (usually calcareous). Their numbers may have declined as these preferred soil types have been replaced by clay subsoil resulting from drainage and cultivation.

2.6.3 Secondary Calcium Limitation

Calcium limitation does not always occur as a direct result of reductions in calcium availability. In heavily acidified areas where toxic cations, such as aluminum and cadmium, accumulate, calcium uptake mechanisms of birds may be severely disrupted

(Eeva and Lehikoinen 1995; Scheuhammer 1991). Toxic cations accumulate in avian tissues, damaging them and disrupting calcium uptake across gut walls. At high levels of dietary calcium intake, ingestion of such cations does not have a toxic effect (Scheuhammer 1987), but, when calcium intake is low, toxic cations can outcompete calcium ions for transport proteins involved in calcium uptake (e.g., Six and Goyer 1970), and birds then show signs of calcium deficiency.

Dichlorodiphenyltrichloroethane (DDT) was discovered in 1939 and became "standard issue" during the Second World War, when it was employed to check the spread of typhus, malaria, and other diseases among the Allied Forces and civilian populations. By the end of the war, so much DDT had been manufactured that it was widely used in veterinary services, agriculture, and forestry (Dunlap 1981). By the 1970s it was used as a biocide on a global scale. DDT readily attacks the nervous systems of insects and is highly effective in controlling their numbers. However, DDT is also readily passed from one animal to another through the food chain. In birds and mammals, DDT is transformed to (1,1'-dichloroethenylidene)-bis-(4-chlorobenzene) (DDE), its main fat-soluble metabolite, which accumulates in the somatic fat reserves and in egg yolk. Although not as toxic as DDT, DDE degrades more slowly and has severe detrimental effects on the metabolism of birds. These result in declines in the breeding performance of birds associated with exposure to DDT, as has been well-documented for many species, particularly those at higher trophic levels (Hickey and Anderson 1968). These include the Peregrine Falcon (*Falco peregrinus)* (Newton et al. 1989; Ratcliffe 1970), the Eurasian Sparrowhawk (*Accipiter nisus*) (Newton and Bogan 1974), the Black-crowned Night-Heron (*Nycticorax nycticorax*) (Ohlendorf et al. 1988; McEwen et al. 1984), and the Double-crested Cormorant (*Phalacrocorax auritus*) (Weseloh et al. 1995; Gress et al. 1973). At least some of this lowering of breeding performance may be attributable to the effect of DDT on calcium metabolism.

2.6.3.1 Disruption of Reproduction by DDT

Accumulation of DDT (and DDE) in avian tissues results in thinning of the eggshell. Cooke (1975) compared eggshells of Domestic Chickens on low dietary calcium with those of birds on chemicals that reduce calcium deposition, with Mallards treated with DDT, and with Grey Herons (*Ardea cinerea*) known to have been exposed to organochlorine insecticides. While the mammillary and palisade layers, the main layers of the eggshell, were of similar thickness for Domestic Chickens and herons, some parts of the thin shells produced by DDT-treated ducks were similar in thickness to eggs produced by untreated birds. This suggested that DDT-induced defects in shell structure resulted from a premature cessation of growth of the calcified layers of the eggshell, rather than from a shortage of shell components. This is further supported by a study of eggs of Eurasian Sparrowhawks and Peregrine Falcons comparing eggs taken before the use of DDT with contemporary eggs. Cooke (1979) found that the mammillary and palisade layers in contemporary eggs underwent roughly proportional reductions in thickness. In addition to disruption of eggshell calcification, Jefferies (1967) found that DDT also caused infertility (delays in ovulation), reductions in hatching and fledging success, extensions of incubation and chick-rearing periods, reductions in body weight, and death of neonates (see Jones and Summers 1968).

Disruption of eggshell calcification by DDT is mediated through the inhibition of prostaglandin synthesis (Lundholm 1997). Concentrations of prostaglandins greatly increase 6–8 h after ovulation and remain high 6–16 h after ovulation, during the period of eggshell calcification (Hertlendy 1980). Prostaglandins are powerful stimulators of uterine contraction and, more specifically, prostaglandin E_2 influences the transport of calcium ions across the shell gland during eggshell formation. DDE inhibits the activity of prostaglandin synthetase, reducing the production of prostaglandin E_2 by the mucosa of the shell gland and reducing uptake of calcium during shell formation (Lundholm, 1994; 1997). As a result, Lundholm (1997) found that the concentration of calcium ions was significantly reduced in the lumen of the shell gland during shell formation, and shell thinning of eggs occurred as a result.

2.6.3.2   Is DDT Responsible for Recent Declines in Breeding Performance of Birds?

DDT was restricted in its use in Britain in 1964 and in the USA in 1972, but in other areas, such as Southeast Asia and Central America, DDT is still widely used. Over 25 years later, the adverse effects of DDT on bird populations persist in North America and these have been particularly apparent in fish-eating birds (Grasman et al. 1998). Comparisons of Double-crested Cormorant eggs measured pre- and post-DDT exposure demonstrate a significant reduction in eggshell thickness (Weseloh et al. 1995; Gress et al. 1973). Weseloh et al. (1995) attributed the decline in cormorant numbers on the Great Lakes, from ~900 pairs in 1950 to virtual extinction by the early 1970s, to DDE-induced shell thinning. Since DDT was banned, the population has expanded by 29% per annum. Similarly, Audet et al. (1992) found that DDE residues declined significantly in Osprey (*Pandion haliaetus*) eggs analyzed in the mid-1980s compared with those from the early 1970s. However, long-term trends in eggshell thickness of shorebirds (Morrison and Kiff 1979) and American White Pelicans (*Pelecanus erythrorhynchos*) (Budgen and Evans 1997) show little change over the decades in which DDT was employed.

In Europe, raptorial birds have suffered reproductive failure and population declines that can be attributed to the accumulation of DDT. Perhaps the best-documented decline is that of the Peregrine Falcon in Britain (Mellanby 1992), where it suffered a high incidence of eggshell thinning and breakage during incubation. Shell breakage occurred in ~4% of clutches prior to 1939 but rose to ~39% after 1951. Shell thickness was constant for Peregrine eggs measured from 1850 to 1947 but then decreased rapidly in 1947. The decline in numbers of Peregrines was not uniform across Britain but was concentrated in those areas where DDT was commonly applied. With the ban on DDT, DDE levels in tissues and eggs declined and Peregrine populations recovered across extensive areas of the UK and western Europe (Newton et al. 1989). Similar to the situation in North America, European seabirds and shorebirds have not experienced significant reductions in eggshell thickness over the years that DDT was employed extensively (e.g., Pulliainen and Marjakangas 1980).

DDT continues to reduce reproductive success of birds in areas where its widespread usage persisted up until a few years ago. For example, in Zimbabwe DDT was banned from garden and domestic use in 1973 and from agricultural use in 1985, but it was still used to control Tsetse Fly (*Glossina* spp.) until the end of 1995. DDE contamination of African Fish Eagles (*Haliaeetus vocifer*) and African Goshawks (*Accipiter tachiro*) has mirrored the frequency of DDT spraying (Douthwaite 1992), particularly in the Zambezi Valley where Tsetse Fly control was most intense. Measurements of eggshell thickness of Peregrine Falcons in Zimbabwe have demonstrated mean shell thickness to be 10% lower than the pre-DDT average (Hartley et al. 1995).

2.7 Calcium Supplementation

Calcium supplementation can be implemented only over a restricted area and does not result in enduring improvements in calcium availability; nevertheless, the effects of elevated calcium availability on breeding performance of birds can be better understood in studies in which calcium supplements have been provided. Table 2.3 lists the effects of calcium supplementation on the breeding performance of various species. The most dramatic results were those of Graveland et al. (1994) and Graveland and Drent (1997) in The Netherlands. Birds were nesting in an area of poorly buffered, acidified podzolic soil on which the abundance of small snails (their main source of calcium during laying) had suffered a long-term decline. Calcium supplementation with fragments of snail shell and eggshell produced marked improvements in many reproductive parameters (Table 2.3). Where birds nest in areas of good soil quality and where acid precipitation is low, calcium supplementation may not significantly improve their breeding performance. Such was the case with House Wrens (*Troglodytes aedon*) nesting in Wyoming (Johnson and Barclay 1996) in an area where the soil, a Quaternary alluvium, consisted of eroded carbonates of calcium and magnesium, sand, and shale, and had a high buffering capacity. Birds whose diets were supplemented with crushed oyster shell and eggshell fragments showed no significant improvements in any reproductive parameters, suggesting that they were obtaining sufficient calcium for egg formation without calcium supplements.

Even in areas where birds would seemingly benefit from increases in calcium availability, calcium supplementation sometimes has no effect on breeding performance. For instance, nesting Blue Tits were provided with additional calcium (Ramsay and Houston 1999) in an area near Loch Lomond, whose soil was of lower quality than that of the Buunderkamp forest in The Netherlands (Graveland et al. 1994).

Table 2.3 Effects of calcium supplementation on reproductive parameters of birds. Adapted from Reynolds et al. (2004)

Species	Effects	Reference
Cape Vulture (*Gyps coprotheres*)	Decrease in incidents of osteodystrophy in chicks	Richardson et al. 1986
Black Tern (*Chlidonias niger*)	Increase in body mass of 15-day-old chicks	Beintema et al. 1997
Tree Swallow (*Tachycineta bicolor*)	No effect on chick survival but positive effect on growth rate	Dawson and Bidwell 2005
Great Tit (*Parus major*)	Reductions in number of females without eggs, clutch desertion, defective eggshells, and nonhatched eggs per clutch. No effect on clutch size or laying date. Positive effect on clutch size, egg volume, egg yolk calcium concentration, eggshell mass and thickness, lay date, brood size, chick growth, and fledgling numbers	Graveland and Drent 1997; Graveland et al. 1994; Tilgar 2002; Tilgar et al. 1999a, b, 2002, 2005; Mänd et al. 1998, 2000
Blue Tit (*Cyanistes caeruleus*)	No effect on egg mass and volume, shell mass, shell thickness index, onset of laying, clutch size, or fledging success	Ramsay and Houston 1999
European Pied Flycatcher (*Ficedula hypoleuca*)	Positive effect on egg volume, eggshell mass and thickness, lay date chick growth, and female condition	Mänd and Tilgar 2003; Tilgar 2002; Tilgar et al. 1999a, b; Mänd et al. 1998
Purple Martin (*Progne subis*)	No effect on growth rate of nestlings	Poulin and Brigham 2001
House Wren (*Troglodytes aedon*)	No effect on egg size, number of fledglings or fledgling body mass. Tendency to lay more eggs	Johnson and Barclay 1996

Supplements of cuttlefish bone, oyster grit, and crushed eggshell resulted in no significant changes in reproductive parameters of birds. These results suggested that despite low snail availability, Blue Tits in the Loch Lomond area were not as calcium-stressed as Great Tits in The Netherlands, which showed high rates of eggshell defects.

Physical and chemical properties of calcium supplements can affect breeding performance. Crushed snail shell, oyster shell, and chicken eggshell provide digestible sources of calcium for many passerines, and providing those in areas where birds are severely calcium-limited can produce rapid improvements in breeding performance (e.g., Graveland et al. 1994). Calcareous grit is also consumed by many species during laying, probably both as a grinding agent and as a source of dietary calcium. Bendell-Young and Bendell (1999) found that grit ingested by Spruce Grouse females could contribute up to 30% of the total calcium content of the gizzard. Although grit therefore provides essential nutrients to the laying bird (e.g., calcium, copper, iron, phosphorus), they also found that grit ingestion resulted in exposure to toxic cations, most notably cadmium. Analysis of liver contents indicated that cadmium accumulated with age. Little is known about grit turnover times in birds and correspondingly little about the resultant exposure of birds to toxic metals through grit ingestion. If grit turnover by laying birds is particularly high, the elemental composition of supplemented calcareous grits should be considered before providing them as calcium supplements.

Reproductive success of some species can depend upon the availability of calcium supplements in relation to the timing of egg formation, embryogenesis, and chick-rearing. In addition to the period of shell formation, some species require calcium-rich foods that they feed directly to their chicks for mineralization of their skeletons. Failure by parents to provide sufficient calcium-rich foods can cause chicks to develop rickets (Fig. 2.4). For example, the low calcium content of the soft tissues of ungulates, which are the major dietary component of seven species of *Gyps* vultures, has probably caused these birds to lay a single egg and to care for chicks over a protracted period (Houston 1978). Vultures face dietary problems even during the nonbreeding season because they must supplement their diets with calcium-rich foods if they are to prevent calcium deficiency (e.g., Bertran and Margalida 1997). Furthermore, to breed successfully, adults must supply their chicks with easily digestible calcium-rich material, usually small bone fragments. Some vultures are found in areas where the only source of calcium is bone fragments resulting from carnivores crushing the bones of their prey. Richardson et al. (1986) found that in ranched areas of South Africa, where Spotted Hyenas (*Crocuta crocuta*) were absent, Cape Vultures provided their chicks with large unfragmented bones. As a result, many chicks developed rickets whereas in wild areas, where hyena-produced bone fragments were abundant, none of the 387 chicks showed signs of rickets. Richardson et al. (1986) provided bone "restaurants" (crushed skeletons) at artificial feeding sites and reduced the incidence of rachitic chicks from 16.9% in 1976 to 2–5% in 1983. Through the success of this supplementary feeding program, metabolic bone disease continue to be rare in vultures and "restaurants," providing fresh food and bone fragments on a regular basis, are a valuable vulture conservation tool (e.g., Verdoorn 1998).

A fuller account of calcium supplementation and how it can be used to study calcium-limited reproduction in birds is provided by Reynolds et al. (2004).

2.8 Future Research

Calcium-specific foraging by laying birds has been reported for many species for which calcium derived from the skeleton is insufficient to meet the demands for producing even one egg (e.g., Campbell and Leatherland 1983). In addition, many birds provide calcium-rich material to their chicks immediately after they have hatched. Ensuring high availability of calcium to birds during breeding, at least to those that do not routinely consume high calcium diets at other times of the year, is fundamental.

Although some cases of calcium-limited reproduction have been reported, many species may already be suffering from this limitation, but exhibiting declines in reproductive success that are too subtle to detect. It is only when calcium availability drops below a dietary threshold for any given species that declines in breeding performance are dramatic and therefore obvious. Knowledge of the dietary calcium requirements of laying poultry species permits poultry scientists to sustain high egg production by providing well-formulated feeding regimes in which calcium availability is monitored closely. Such detailed knowledge for nonpoultry species is lacking, but research to date indicates that most birds rely upon exogenous, as well as endogenous, dietary calcium for calcification of the eggshell. Furthermore, regardless of the time of laying (e.g., Domestic Chicken [morning] vs. Common Pigeon [afternoon]), it appears that some dietary calcium is deposited in short-term medullary bone reserves to be mobilized for eggshell formation at times when dietary intake of calcium is low. Aspects of calcium physiology may be shared between laying Domestic Chickens and other species, and the poultry literature may be an invaluable resource in helping us to understand more about the calcium logistics of reproduction in all bird species.

2.8.1 Basic Dietary Information

Although a growing number of avian species have been observed feeding on calcium-rich food (e.g., Dhondt and Hochachka 2001) items during laying, there remains a large number of species for which our knowledge of breeding diet is extremely limited. Detailed studies of foraging behavior would demonstrate the range of calcareous material that is exploited by breeding birds. For instance, Graveland (1996a) found that, unlike Great Tits, European Pied Flycatchers in the same forest did not suffer from calcium deficiency. Millipedes and woodlice were consumed as calcium sources during laying by European Pied Flycatchers, which are thin-billed insectivores that find it difficult to process fragments of snail shell. Instead, they obtain sufficient calcium for egg production by consuming arthropods with calcified exoskeletons. Therefore, it is surprising that supplementation of European Pied Flycatchers' diet with snail and eggshell fragments resulted in marked improvements in egg and laying parameters (Table 2.3, Tilgar et al. 1999a; Mänd et al. 1998). Omnivorous birds may switch diet during breeding when preferred calcium-rich foods become scarce.

Although eggshell is rich in calcium, most altricial birds remove all fragments of eggshell from the nest soon after hatching. The white lining of shells is highly conspicuous to predators and shell removal may reduce the risk of predation of chicks and unhatched eggs (Tinbergen et al. 1962). However, in species that have concealed nests, predation pressure is usually sufficiently low that other explanations need to be sought. These may include prevention of bacterial infection of material adhered to shell fragments, injuries to naked chicks from sharp shell fragments, disturbance of brooding adults, and "egg-capping" (Derrickson and Warkentin 1991), where shells slip over unhatched eggs and interfere with the pipping process and gaseous exchange across eggshells. Few birds appear to exploit readily accessible calcium-rich eggshell by consuming it themselves or feeding it to their nestlings (but see Löhrl 1978). Verbeek (1996) noted that only 25 out of the 160 first-published species (excluding two cowbird *Molothrus* spp.) described in "The Birds of North America" series leave eggshells in the nest after chicks have hatched. Only 20 of these either routinely eat eggshells or do so in combination with other methods of disposal (i.e., carrying them off or trampling them into the nest). Further intensive study of birds at the nest may reveal a greater prevalence of eggshell consumption by both adults and chicks.

Extensive areas of northern Europe, the Canadian Shield, and Africa are composed of gneiss or granite, noncalcareous rocks supporting few sources of calcium-rich food. As a result, in North America some species are less abundant in calcium-poor areas (e.g., Common Pheasant, Dale 1954; Grey Partridge [*Perdix perdix*], Wilson 1959), and birds that are present may invest much time and energy foraging for calcium during laying. Tilgar et al. (1999a, b) suggested that in years when food availability was low, Great Tits and European Pied Flycatchers, feeding in base-poor areas of Estonia, struggled to obtain sufficient calcium for egg formation and postponed egg laying as a result. Under such conditions, these authors propose that laying birds incur extra costs, over and above those estimated on the basis of energetic and nutrient requirements. In particularly base-poor areas, laying birds may spend significantly more time foraging for calcium-rich material and correspondingly less time searching for other nutrients. Until we have information about soil quality (e.g., pH, cation exchange capacity, buffering capacity), pollution levels, and historical land-use patterns of these areas, it is impossible to know whether availability of appropriate calcium-rich foods is stable or declining at a given site. Reliable predictions about the stability of avian populations within these areas will be possible only when such data are available, together with accurate bird census data (Graveland 1998). Even then, relating avian distributions to calcium availability can be seriously confounded by other factors, such as the availability of other nutrients (see Gionfriddo and Best 1999) and of suitable nest sites. Patten (2007) examined the proximate and ultimate effects of dietary calcium on clutch size and, in so doing, he has taken the first steps in understanding how broad-scale calcium availability might explain some of the patterns in calcium-limited life-history parameters such as clutch and brood size.

2.8.2 Supplementation of Birds' Diets During Breeding

Few birds nesting in areas of low calcium availability have been given dietary supplements of calcium-rich food, and those few studies (Table 2.3) have produced equivocal results. Further field studies in which birds' diets are supplemented with calcium will substantially increase our understanding of the timing of calcium intake and the importance of calcareous material to successful egg formation and chick development. Such research might be imperative if some critically endangered species are to survive. For example, breeding success of translocated populations of the Kakapo (*Strigops habroptilus*) has been ameliorated through supplementary feeding of protein-rich foods (Elliott et al. 2001), but negative effects of supplementation (e.g., obesity of breeding females, Powlesland and Lloyd 1994) have necessitated a reconsideration of supplementation protocols. Raubenheimer and Simpson (2006) suggested that calcium might be limiting in Kakapo reproduction and, if this be the case, that reformulation of supplementary diets to lower the macronutrient: calcium ratio might hold the key to improving their reproductive output.

2.8.3 Calcium Uptake and Utilization

Learning about the physiological mechanisms involved in the uptake and utilization of calcium during egg laying will assist in pinpointing the stages of reproduction at which calcium is most critical. Laboratory investigations using drugs that restrict the availability of calcium during egg laying would be highly informative. For example, Cooke (1975) administered sulfanilamide to laying Domestic Chickens and found that both mammillary and palisade layers of the eggshell were reduced in thickness compared with those laid by control (nondrugged) birds. However, the egg was retained for a normal time in the shell gland and was laid with an intact cuticle. These findings suggest that sulfanilamide disrupts the supply of carbonate and calcium ions to the shell gland, resulting in a slower rate of calcium deposition, but it has no significant effect on the schedule of the calcification process itself. Further research to investigate the mechanism of action of sulfanilamide would allow such drugs to be used in tightly controlled feeding trials with captive birds. Such calcium antagonists/inhibitors have the potential to enhance our understanding of the calcium logistics of birds during reproduction.

Feeding diets of different calcium content to captives would allow identification of a dietary calcium threshold below which birds suffer severe limitation. Eggs could be readily collected from birds whose dietary calcium intake is known and ultrastructural abnormalities in shell structure could be examined. Currently, little is known about the effects of subtle changes in dietary calcium availability on eggshell structure, or about how structural defects in eggshells ultimately affect the breeding performance of wild birds.

In the field, targeted studies allow invaluable insights into the timing of calcium intake in relation to egg laying and chick-rearing (see Bureš and Weidinger 2001) and biochemical markers targeting skeletal development are now being used to examine the timing of mineralization during skeletal development (e.g., Tilgar et al. 2005, 2008). A further strand of research might involve free-living species that ingest food containing negligible amounts of calcium. Hummingbirds, for instance, require minerals, especially during reproduction, that are found in nectar in inadequate quantities to meet even some metabolic requirements (Hainsworth and Wolf 1976). They appear to meet their daily energy requirements by consuming insects as well as nectar (Remsen et al. 1986). Their needs beyond those for self-maintenance (e.g., calcium needed for egg production) are met by supplementing their diets during and after reproduction with calcium-rich soil and ash (Adam and des Lauriers 1998). Providing calcium supplements to hummingbirds would allow us to understand the calcium physiology of a group of organisms that do not routinely consume calcium-rich food.

It is difficult to know to what extent the results from captive studies can be applied to free-living individuals (Lambrechts et al. 1999). Klasing (1998) questions whether domesticated animals have similar mineral requirements to free-living, nondomesticated ones, the former having been selected for rapid growth and production. Although earlier in this chapter we advocated increased use of the poultry literature in order to enhance our knowledge of calcium logistics of laying birds, in general, caution should be exercised. The Domestic Chicken has undergone more than 50 years of selection for high egg production, and its mechanisms of calcium uptake and utilization may not be representative of other avian species during eggshell formation. Other captive species are often maintained on a surfeit of foods that can have large effects on the efficiency of nutrient utilization and even sometimes on gut morphology. Even in the Willow Ptarmigan (*Lagopus lagopus*), for example, not only do captive males have much shorter cecal and intestinal lengths (Moss 1972) than wild males, but they also have much lower digestive efficiencies (27 and 46%, respectively) (Moss 1977).

Despite these reservations about the applicability of studies of domesticated and captive species to wild birds, future research should combine laboratory and field-based studies. Investigations (e.g., Pahl et al. 1997; Reynolds 1997) have certainly suggested some parallels between small passerines and domesticated poultry species in terms of their calcium physiology during egg formation. Furthermore, Blom and Lilja (2004) used a comparative approach to demonstrate intriguing relationships between eggshell microstructure and growth patterns of both poultry and free-living, nondomesticated species that significantly advance our understanding of avian life-history patterns. Poultry science may contribute much to our currently limited knowledge of the calcium requirements for successful reproduction in wild birds.

Acknowledgments We would like to thank Val Nolan and Charlie Thompson for their help during preparation of this chapter. We also thank Ian Carter of Natural England and Sue Thornton of the Institute of Zoology at the Zoological Society of London for allowing us to reproduce X-rays of Red Kites. Kaat Brulez, Phill Cassey, Paul Donald, Kim Fernie, Ian Hill, Andy Radford, Sophie Stafford, and Jeremy Wilson provided numerous suggestions that helped to improve the chapter. We are also grateful to members of the Editorial Board of "Current Ornithology" for their constructive and diverse comments on the prospectus. Much of this work was made possible through financial support from the NERC to SJR.

References

Adam MD, des Lauriers JR (1998) Observations of hummingbirds ingesting mineral-rich compounds. J Field Ornithol 69:257–261
Alisauskas RT, Ankney CD (1994) Nutrition of breeding female Ruddy Ducks: the role of nutrient reserves. Condor 96:878–897

Alisauskas RT, Eberhardt RT, Ankney CD (1990) Nutrient reserves of breeding Ring-necked Ducks (*Aythya collaris*). Can J Zool 68:2524–2530

Ammerman CB, Baker DH, Lewis AJ (1995) Bioavailability of nutrients for animals: amino acids, minerals, and vitamins. Academic Press, Oxford

Anderson TR (1995) Removal indeterminacy and the proximate determination of clutch size in the House Sparrow. Condor 97:197–207

Ankney CD, Afton AD (1988) Bioenergetics of breeding Northern Shovelers: diet, nutrient reserves, clutch size and incubation. Condor 90:459–471

Ankney CD, Alisauskas RT (1991) Nutrient-reserve dynamics and diet of breeding female Gadwalls. Condor 93:799–810

Ankney CD, MacInnes CD (1978) Nutrient reserves and reproductive performance of female Lesser Snow Geese. Auk 95:459–471

Ankney CD, Scott DM (1980) Changes in nutrient reserves and diet of breeding Brown-headed Cowbirds. Auk 97:684–696

Annett C, Pierotti R (1989) Chick hatching as a trigger for dietary switching in the Western Gull. Colonial Waterbirds 12:4–11

Ar A, Rahn H (1980) Water in the avian egg: overall budget of incubation. Am Zool 20: 373–384

Ar A, Paganelli CV, Reeves RB, Greene DG, Rahn H (1974) The avian egg: water vapor conductance, shell thickness, and functional pore area. Condor 76:153–158

Arnold TD, Ankney CD (1997) The adaptive significance of nutrient reserves to breeding American Coots: a reassessment. Condor 99:91–103

Astheimer LB, Grau CR (1985) The timing and energetic consequences of egg formation in the Adélie Penguin. Condor 87:256–268

Audet DJ, Scott DS, Wiemeyer SN (1992) Organochlorines and mercury in Osprey eggs from the eastern United States. J Raptor Res 26:219–224

Baimbridge KG, Taylor TG (1981) The role of calcitonin in controlling hypercalcemia in the domestic fowl (*Gallus domesticus*). Comp Biochem Physiol A 68:647–651

Bairlein F, Simons D (1995) Nutritional adaptations in migrating birds. Isr J Zool 41:357–367

Balkan M, Karaka R, Biricik M (2006) Changes in eggshell thickness, shell conductance and pore density during incubation in the Peking Duck (*Anas platyrhynchos* f. *dom.*). Ornis Fenn 83:117–123

Barrentine CD (1980) The ingestion of grit by nestling Barn Swallows. J Field Ornithol 51:368–371

Baumgartner S, Brown DJ, Salevsky E, Leach RM (1978) Copper deficiency in the laying hen. J Nutr 108:804–811

Beasom SL, Pattee OH (1978) Utilization of snails by Rio Grand Turkey hens. J Wildl Manage 42:916–919

Beintema AJ, Baarspul T, de Krijger JP (1997) Calcium deficiency in Black Terns *Chlidonias niger* nesting on acid bogs. Ibis 139:396–397

Bendell-Young LI, Bendell JF (1999) Grit ingestion as a source of metal exposure in the Spruce Grouse, *Dendragapus canadensis*. Environ Pollut 106:405–412

Bertran J, Margalida A (1997) Griffon Vultures (*Gyps fulvus*) ingesting bones at the ossuaries of Bearded Vultures (*Gypaetus barbatus*). J Raptor Res 31:287–288

Betts MM (1955) The food of titmice in oak woodland. J Anim Ecol 24:282–323

Biebach H (1992) Flight-range estimates for small trans-Sahara migrants. Ibis 134(suppl 1):47–54

Bilby LW, Widdowson EM (1971) Chemical composition of growth in nestling Blackbirds and thrushes. Br J Nutr 25:127–134

Blancher PJ, McAuley DG (1987) Influence of wetland acidity on avian breeding success. Trans N Am Wildl Nat Resour Conf 52:628–635

Blancher PJ, McNicol DK (1988) Breeding biology of Tree Swallows in relation to wetland acidity. Can J Zool 66:842–849

Blancher PJ, McNicol DK (1991) Tree Swallow diet in relation to wetland acidity. Can J Zool 69:2629–2637

Blom J, Lilja C (2004) A comparative study of growth, skeletal development and eggshell composition in some species of birds. J Zool Lond 262:361–369

Bloom MA, Domm LV, Nalbandov AV, Bloom W (1958) Medullary bone of laying chickens. Am J Anat 102:411–453

Blum JD, Taliaferro EH, Holmes RT (2001) Determining the sources of calcium for migratory songbirds using stable strontium isotopes. Oecologia 126:569–574

Booth DT, Seymour RS (1987) Effect of eggshell thinning on water vapour conductance of Malleefowl (*Laipoa ocellata*) eggs. Condor 89:453–459

Brenninkmeijer A, Klaassen M, Stienen EWM (1997) Sandwich Terns *Sterna sandvicensis* feeding on shell fractions. Ibis 139:397–400

Budgen SC, Evans RM (1997) Egg composition and post-DDT eggshell thickness of the American White Pelican, *Pelecanus erythrorhynchos*, Can Field-Nat 111:234–237

Bump G, Darrow RW, Edminster FC, Crissey WF (1947) The Ruffed Grouse: Life history, propagation and management. New York Conservation Department, Albany, New York

Bureš S, Weidinger K (2001) Do pipits use experimentally supplemented rich sources of calcium more often in acidified areas? J Avian Biol 32:194–198

Bureš S, Weidinger K (2003) Sources and timing of calcium intake during reproduction in flycatchers. Oecologia 137:634–647

Burley RW, Vadehra DV (1989) The avian egg: chemistry and biology. Wiley-Interscience, New York

Burmester BR (1940) A study of the physical and chemical changes of the egg during its passage through the isthmus and uterus of the hen's oviduct. J Exp Zool 84:445–500

Byrkjedal I (1975) Skeletal remains of microrodents as a source of calcium for Golden Plovers, *Pluvialis apricaria*, during the egg laying period. Sterna 14:197–198

Campbell RR, Leatherland JF (1983) Changes in calcium reserves in breeding Lesser Snow Geese (*Chen caerulescens caerulescens*). Acta zool (Stockh) 64:9–14

Carlsson H, Carlsson L, Wallin C, Wallin N-E (1991) Great Tits incubating empty nest cups. Ornis Svecica 1:51–53

Chattopadhyay S (1980) Egg-bound death of a Purplerumped Sunbird at Baj Baj, West Bengal. J Bombay Nat Hist Soc 77:333

Clunies M, Etches RJ, Fair C, Leeson S (1993) Blood, intestinal and skeletal calcium dynamics during egg formation. Can J Anim Sci 73:517–532

Cooke AS (1975) Pesticides and eggshell formation. Symp Zool Soc Lond 35:339–361

Cooke AS (1979) Changes in egg shell characteristics of the Sparrowhawk and Peregrine associated with exposure to environmental pollutants during recent decades. J Zool (Lond) 187:245–263

Dacke CG, Arkle S, Cook DJ, Wormstone IM, Jones S, Zaidi M, Bascal ZA (1993) Medullary bone and avian calcium regulation. J Exp Biol 184:63–88

Dale FH (1954) Influence of calcium on the distribution of the Pheasant in North America. Trans N Am Wildl Nat Resour Conf 19:316–323

Dalke PD (1938) Amount of grit taken by pheasants in southern Michigan. J Wildl Manage 2:53–54

Davies NB (1977) Prey selection and the search strategy of the Spotted Flycatcher (*Muscicapa striata*): a field study on optimal foraging. Anim Behav 25:1016–1033

Dawson RD, Bidwell MT (2005) Dietary calcium limits size and growth of nestling tree swallows *Tachycineta bicolor* in a non-acidified landscape. J Avian Biol 36:127–134

Dean R (1989) Pearlbreasted Swallow apparently eating lime. Promerops 189:13–14

Dekhuijzen HM, Schuijl GPJ (1996) Veranderingen in het broedsucces van Koolmees *Parus major* en Bonte Vliegenvanger *Ficedula hypoleuca* op de Veluwe en in het Gooi van 1973-92. Limosa 69:165–174

Derrickson KC, Warkentin IG (1991) The role of egg-capping in the evolution of eggshell removal. Condor 93:757–759

DesGranges J-L, Mauffette Y, Gagnon G (1987) Sugar maple forest decline and implications for forest insects and birds. Trans N Am Wildl Nat Resour Conf 52:677–689

Dhondt AA, Hochachka WM (2001) Variations in calcium use by birds during the breeding season. Condor 103:592–598

Diamond J, Bishop KD, Gilardi JD (1999) Geophagy in New Guinea birds. Ibis 141:181–193

Douthwaite RJ (1992) Effects of DDT on the Fish Eagle *Haliaeetus vocifer* population of Lake Kariba in Zimbabwe. Ibis 134:250–258

Drent RH, Daan S (1980) The prudent parent: energetic adjustments in avian breeding. Ardea 68:225–252

Drent PJ, Woldendorp JW (1989) Acid rain and eggshells. Nature 339:431

Drobney RD (1980) Reproductive bioenergetics of Wood Ducks. Auk 97:480–490

Dunlap TR (1981) DDT: Scientists, citizens and public policy. University Press, Princeton, New Jersey

Dunn BE, Boone MA (1977) Growth and mineral content of cultured chick embryos. Poult Sci 57:370–377

Eeva T, Lehikoinen E (1995) Egg shell quality, clutch size and hatching success of the Great Tit (*Parus major*) and the Pied Flycatcher (*Ficedula hypoleuca*) in an air pollution gradient. Oecologia 102:312–323

Elliott GP, Merton DV, Jansen PW (2001) Intensive management of a critically endangered species: the Kakapo. Biol Conserv 99:121–133

Etches RJ (1987) Calcium logistics in the laying hen. J Nutr 117:619–628

Etches RJ, Schoch JP (1984) A mathematical representation of the ovulatory cycle of the domestic hen. Br Poult Sci 25:65–76

Ficken MS (1989) Boreal Chickadees eat ash rich in calcium. Wilson Bull 101:349–351

Fogden MPL, Fogden PM (1979) The role of fat and protein reserves in the annual cycle of the Grey-backed Camaroptera in Uganda (Aves: Sylvidae). J Zool (Lond) 189:233–258

Foster MS (2007) The potential of fruit trees to enhance converted habitats for migrating birds in southern Mexico. Bird Conserv Int 17:45–61

Gibb JA, Betts MM (1963) Food and food supply of nestling tits (Paridae) in breckland pine. J Anim Ecol 32:489–533

Gilardi JD, Duffey SS, Munn CA, Tell LA (1999) Biochemical functions of geophagy in parrots: detoxification of dietary toxins and cytoprotective effects. J Chem Ecol 25:897–922

Gilbert AB (1983) Calcium and reproductive function in the hen. Proc Nutr Soc 42:195–212

Gillings S, Watts PN (1997) Habitat selection and breeding ecology of Corn Buntings *Miliaria calandra* in the Lincolnshire Fens. In: Donald PF, Aebischer NJ (eds) The ecology and conservation of Corn Buntings *Miliaria calandra*. Joint Nature Conservation Committee, Peterborough, pp 139–150

Gionfriddo JP, Best LB (1996) Grit-use patterns in North American birds: the influence of diet, body size, and gender. Wilson Bull 108:685–696

Gionfriddo JP, Best LB (1999) Grit use by birds: a review. In: Nolan V Jr, Ketterson ED, Thompson CF (eds) Current Ornithology, vol 15. Kluwer/Plenum, New York, pp 89–148

Glooschenko V, Blancher P, Herskowitz J, Fulthorpe R, Rang S (1986) Association of wetland acidity with reproductive parameters and insect prey of the Eastern Kingbird (*Tyrannus tyrannus*) near Sudbury, Ontario. Water Air Soil Pollut 30:553–567

Golet GH, Kuletz KJ, Roby DD, Irons DB (2000) Adult prey choice affects chick growth and reproductive success in Pigeon Guillemots. Auk 117:82–91

Gosler AG, Barnett PR, Reynolds SJ (2000) Inheritance and variation in eggshell patterning in the great tit *Parus major*. Proc R Soc Lond B 267:2469–2473

Gosler AG, Higham JP, Reynolds SJ (2005) Why are birds' eggs speckled? Ecol Lett 8:1105–1113

Grasman KA, Scanlon PF, Fox GA (1998) Reproductive and physiological effects of environmental contaminants in fish-eating birds of the Great Lakes: a review of historical trends. Environ Monit Assess 53:117–145

Graveland J (1995) The quest for calcium: calcium limitation in the reproduction of forest passerines in relation to snail abundance and soil acidification. Ph.D. thesis, University of Groningen

Graveland J (1996a) Avian eggshell formation in calcium-rich and calcium-poor soils: importance of snail shells and anthropogenic calcium sources. Can J Zool 74:1035–1044

Graveland J (1996b) Calcium deficiency in wild birds. Vet Q 18(suppl):S136–S137

Graveland J (1998) Effects of acid rain on bird populations. Environ Rev 6:41–54

Graveland J, Berends JE (1997) Timing of the calcium uptake and effect of calcium deficiency on behaviour and egg-laying in captive Great Tits, *Parus major*. Physiol Zool 70:74–84

Graveland J, Drent RH (1997) Calcium availability limits breeding success of passerines on poor soils. J Anim Ecol 66:279–288

Graveland J, van Gijzen T (1994) Arthropods and seeds are not sufficient as calcium sources for shell formation and skeletal growth in passerines. Ardea 82:299–314

Graveland J, van der Wal R, van Balen JH, van Noordwijk AJ (1994) Poor reproduction in forest passerines from decline of snail abundance on acidified soils. Nature 368:446–448

Gress F, Risebrough RW, Anderson DW, Kiff LF, Jehl JR Jr (1973) Reproductive failures of Double-crested Cormorants in southern California and Baja California. Wilson Bull 85:197–208

Hågvar S, Østbye E (1976) Food habits of the Meadow-Pipit, *Anthus pratensis* (L.), in alpine habitats at Hardangervidda, south Norway. Norw J Zool 24:53–64

Hails CJ, Turner AK (1985) The role of fat and protein during breeding in the White-bellied Swiftlet (*Collocalia esculenta*). J Zool (Lond) 206:469–484

Haines TA, Baker JP (1986) Evidence of fish population responses to acidification in the eastern United States. Water Air Soil Pollut 30:605–609

Hainsworth FR, Wolf LL (1976) Nectar characteristics and food selection by hummingbirds. Oecologia 25:101–113

Hames RS, Rosenberg KV, Lowe JD, Barker SE, Dhondt AA (2002) Adverse effects of acid rain on the distribution of the Wood Thrush *Hylocichla mustelina*. Proc Natl Acad Sci USA 99:11235–11240

Hamilton RMG (1982) Methods and factors affecting the measurement of eggshell quality. Poult Sci 61:1192–1197

Harper JA (1964) Calcium in grit consumed by hen pheasants in east-central Illinois. J Wildl Manage 28:264–270

Hartley RR, Newton I, Robertson M (1995) Organochlorine residues and eggshell thinning in the Peregrine Falcon *Falco peregrinus minor* in Zimbabwe. Ostrich 66:69–73

Hertlendy F (1980) Prostaglandins in avian endocrinology. In: Epple A, Stetson MH (eds) Avian endocrinology. Academic Press, New York, pp 455–480

Hickey JJ, Anderson DW (1968) Chlorinated hydrocarbons and eggshell changes in raptorial and fish-eating birds. Science 162:271–273

Higham JP, Gosler AG (2006) Speckled eggs: water-loss and incubation behaviour in the great tit *Parus major*. Oecologia 149:561–570

Hildrew AG, Ormerod SJ (1995) Acidification: causes, consequences and solutions. In: Harper DM, Ferguson AJD (eds) The ecological basis for river management. Wiley, Chichester, pp 147–160

Hirsch PF, Baruch H (2003) Is calcitonin an important physiological substance? Endocrine 21:201–208

Hohman WL (1986) Changes in body weight and body composition of breeding Ring-necked Ducks. Auk 103:181–188

Holmes RT (1994) Black-throated Blue Warbler (*Dendroica caerulescens*). In: Poole A, Gill F (eds) The birds of North America, No. 87. Birds of North America Inc., Philadelphia, Pennsylvania/American Ornithologists' Union, Washington, DC

Houston DC (1978) The effect of food quality on breeding strategy in Griffon Vultures (*Gyps* spp.). J Zool (Lond) 186:175–184

Houston DC, Donnan D, Jones PJ (1995) The source of nutrients required for egg production in Zebra Finches *Poephila guttata*. J Zool (Lond) 235:469–483

Hoyt DF (1979) Osmoregulation by avian embryos: the allantois functions like a toad's bladder. Physiol Zool 52:354–362

Hughes BO (1972) A circadian rhythm of calcium intake by the domestic fowl. Br Poult Sci 13:485–493

Hurwitz S, Bar A (1965) Absorption of calcium and phosphorus along the gastrointestinal tract of the laying fowl as influenced by dietary calcium and egg shell formation. J Nutr 86:433–438

Hurwitz S, Bar A (1969) Intestinal calcium absorption in the laying hen and its importance in calcium homeostasis. Am J Clin Nutr 22:391–395

Jefferies DJ (1967) The delay in ovulation produced by pp'-DDT and its possible significance in the field. Ibis 109:266–272

Jensen LS (1968) Vitamin E and essential fatty acids in avian reproduction. Fed Proc 27:914–919

Johnson AL (2000) Reproduction in the female. In: Whittow GC (ed) Sturkie's avian physiology, 5th edn. Academic Press, London, pp 569–596

Johnson LS, Barclay RMR (1996) Effects of supplemental calcium on the reproductive output of a small passerine bird, the House Wren (*Troglodytes aedon*). Can J Zool 74:278–282

Jones FJS, Summers DDB (1968) Relation between DDT in diets of laying birds and viability of their eggs. Nature 217:1162–1163

Jones PJ (1976) The utilization of calcareous grit by laying *Quelea quelea*. Ibis 118:575–576

Jones PJ, Ward P (1976) The level of reserve protein as the proximate factor controlling the timing of breeding and clutch-size in the Red-billed Quelea *Quelea quelea*. Ibis 118:547–574

Joshua IG, Mueller WJ (1979) The development of a specific appetite for calcium in growing broiler chicks. Br Poult Sci 20:481–490

Karasov WH, Phan D, Diamond JM, Carpenter FL (1986) Food passage and intestinal nutrient absorption in hummingbirds. Auk 103:453–464

Karlsson O, Lilja C (2008) Eggshell structure, mode of development and growth rate in birds. Zoology 111:494–502

Keshavarz K, Nakajima S (1993) Re-evaluation of calcium and phosphorus requirements of laying hens for optimum performance and eggshell quality. Poult Sci 72:144–153

Kilner RM (2006) The evolution of egg colour and patterning in birds. Biol Rev 81:383–406

Klasing KC (1998) Comparative avian nutrition. CAB International, Wallingford

Klaver A (1964) Waarnemingen over de biologie van de Draaihals (*Jynx torquilla* L.). Limosa 37:221–231

Kopischke ED (1966) Selection of calcium- and magnesium-bearing grit by pheasants in Minnesota. J Wildl Manage 30:276–279

Kopischke ED, Nelson MM (1966) Grit availability and pheasant densities in Minnesota and North Dakota. J Wildl Manage 30:269–275

Korschgen CE (1977) Breeding stress of female Eiders in Maine. J Wildl Manage 41:360–373

Korschgen LJ (1964) Foods and nutrition of Missouri and midwestern pheasants. Trans N Am Wildl Nat Resour Conf 29:159–180

Krapu GL (1981) The role of nutrient reserves in Mallard reproduction. Auk 98:29–38

Krementz DG, Ankney CD (1995) Changes in total body calcium and diet of breeding House Sparrows. J Avian Biol 26:162–167

Lambrechts MM, Perret P, Maistre M, Blondel J (1999) Do experiments with captive non-domesticated animals make sense without population field studies? A case study with Blue Tits' breeding time. Proc R Soc Lond B 266:1311–1315

Larison JR, Crock JG, Snow CM (2001) Timing of mineral sequestration in leg bones of White-tailed Ptarmigan. Auk 118:1057–1062

Leach RM Jr (1985) Steel making slag as a source of dietary calcium for the laying hen. Nutr Rep Int 32:475

Löhrl H (1978) Beiträge zur Ethologie und Gewichtsentwicklung beim Wendehals *Jynx torquilla*. Orn Beob 75:193–201

Luck MR, Scanes CG (1982) Calcium homeostasis and the control of ovulation. In: Scanes CG, Ottinger MA, Kenny AD, Balthazart J, Cronshaw J, Chester Jones I (eds) Aspects of avian endocrinology: practical and theoretical implications. Texas Technical University, Lubbock, pp 263–282

Lumeij JT (1990) Relation of plasma calcium to total protein and albumin in African grey (*Psittacus erithacus*) and Amazon (*Amazona* spp.) parrots. Avian Pathol 19:661–667

Lumeij JT (1994) Endocrinology. In: Ritchie BW, Harrison GJ, Harrison LR (eds) Avian medicine: principles and application. Wingers, Lake Worth, pp 582–606

Lundholm CE (1994) Changes in the levels of different ions in the eggshell gland lumen following p, p'-DDE-induced eggshell thinning in ducks. Comp Biochem Physiol C 109:57–62

Lundholm CE (1997) DDE-induced eggshell thinning in birds: effects of p, p'-DDE on the calcium and prostaglandin metabolism of the eggshell gland. Comp Biochem Physiol C 118:113–128

MacLean SF (1974) Lemming bones as a source of calcium for Arctic sandpipers (*Calidris* spp.). Ibis 116:552–557

Mahony N, Nol E, Hutchinson T (1997) Food-chain chemistry, reproductive success, and foraging behaviour of songbirds in acidified maple forests of central Ontario. Can J Zool 75:509–517

Malley DF (1980) Decreased survival and calcium uptake by the Crayfish *Orconectes virilis* in low pH. Can J Fish Aquat Sci 37:364–372

Mänd R, Tilgar V, Leivits A (1998) Calcium deficiency as an ecological constraint for Passerines in oligotrophic forests of Estonia. Biol Conserv Fauna 102:345

Mänd R, Tilgar V, Leivits A (2000) Reproductive response of Great Tits, *Parus major*, to calcium supplementation in a naturally base-poor forest habitat. Can J Zool 78:689–695

Mänd R, Tilgar V (2003) Does supplementary calcium reduce the cost of reproduction in the Pied Flycatcher *Ficedula hypoleuca*. Ibis 145:67–77

March GL, Sadleir RMFS (1975) Studies on the Band-tailed Pigeon (*Columba fasciata*) in British Columbia. VIII. Seasonal changes in body weight and calcium distribution. Physiol Zool 48:49–56

May DL (2001) Grey parrots of the Congo basin forest. PsittaScene 13:8–10

McEwen LC, Stafford CJ, Henser GL (1984) Organochlorine residues in eggs of Black-crowned Night-Herons (*Nycticorax nycticorax*) from Colorado and Wyoming (USA). Environ Toxicol Chem 3:367–376

Meijer T, Drent R (1999) Re-examination of the capital and income dichotomy in breeding birds. Ibis 141:399–414

Mellanby K (1992) The DDT story. British Crop Protection Council, Farnham

Mills KH, Schindler DW (1986) Biological indicators of lake acidification. Water Air Soil Pollut 30:779–789

Morrison ML, Kiff LF (1979) Eggshell thickness in American shorebirds before and since DDT. Can Field-Nat 93:187–190

Moss R (1972) Effects of captivity on gut lengths in Red Grouse. J Wildl Manage 36:99–104

Moss R (1977) The digestion of heather by Red Grouse during the spring. Condor 79:471–477

Mundy PJ, Ledger JA (1976) Griffon Vultures, carnivores and bones. S Afr Sci J 72:106–110

Mundy PJ, Butchart D, Ledger JA, Piper SE (1992) The vultures of Africa. Acorn Books and Russel Friedman Books, Randburg

Murton RK, Westwood NJ, Isaacson AJ (1964) The feeding habits of the Woodpigeon *Columba palumbus*, Stock Dove *C. oenas* and Turtle Dove *Streptopelia turtur*. Ibis 106:174–188

Nager RG, Rüegger C, van Noordwijk AJ (1997) Nutrient or energy limitation on egg formation: a feeding experiment in Great Tits. J Anim Ecol 66:495–507

National Research Council (1984) National requirements of poultry. National Academy Press, Washington, DC

Naylor BJ, Bendell JF (1989) Clutch size and egg size of Spruce Grouse in relation to spring diet, food supply, and endogenous reserves. Can J Zool 67:969–980

Newton I (2008) The migration ecology of birds. Academic, London

Newton I, Bogan J (1974) Organochlorine residues, eggshell thinning and hatching success in British Sparrowhawks. Nature 249:582–583

Newton I, Bogan JA, Haas MB (1989) Organochlorines and mercury in the eggs of British Peregrines. Ibis 131:355–376

Nisbet ICT (1997) Female Common Terns *Sterna hirundo* eating mollusc shells: evidence for calcium deficits during egg laying. Ibis 139:400–401

Nyholm NEI, Myhrberg HE (1977) Severe eggshell defects and impaired reproductive capacity in small passerines in Swedish Lapland. Oikos 29:336–341

O'Brien TG, Kinnaird ME, Dierenfeld ES, Conklin-Brittain NL, Wrangham RW, Silver SC (1998) What's so special about figs? Nature 392:668

Ohlendorf HM, Custer TW, Lowe RW, Rigney M, Cromartie E (1988) Organochlorines and mercury in eggs of coastal terns and herons in California. Colonial Waterbirds 11:85–94

Ormerod SJ, Bull KR, Cummins CP, Tyler SJ, Vickery JA (1988) Egg mass and shell thickness in Dippers *Cinclus cinclus* in relation to stream acidity in Wales and Scotland. Environ Pollut 55:107–121

Ormerod SJ, O'Halloran J, Gribbin SD, Tyler SJ (1991) The ecology of Dippers *Cinclus cinclus* in relation to stream acidity in upland Wales: breeding performance, calcium physiology and nestling growth. J Appl Ecol 28:419–433

Pahl R, Winkler DW, Graveland J, Batterman BW (1997) Songbirds do not create long-term stores of calcium in their leg bones prior to laying: results from high resolution radiography. Proc R Soc Lond B 264:239–244

Patten MA (2007) Geographic variation in calcium and clutch size. J Avian Biol 38:637–643

Payne RB (1972) Nuts, bones, and a nesting of Red Crossbills in the Panamint Mountains California. Condor 74:485–486

Perrins CM, Birkhead TR (1983) Avian ecology. Blackie, Glasgow

Petersen CF (1965) Factors influencing egg shell quality – a review. World Poult Sci 21:110–138

Pierotti R, Annett CA (1987) Reproductive consequences of dietary specialization and switching in an ecological generalist. In: Kamil AC, Krebs JR, Pulliam HR (eds) Foraging behavior. Plenum, New York, pp 417–442

Pierotti R, Annett CA (1991) Diet choice in the Herring Gull: constraints imposed by reproductive and ecological factors. Ecology 72:319–328

Piersma T, Gudmundsson GA, Davidson NC, Morrison RIG (1996) Do Arctic-breeding Red Knots (*Calidris canutus*) accumulate skeletal calcium before egg laying? Can J Zool 74:2257–2261

Pinowska B, Kraśnicki K (1985a) Changes in the content of magnesium, copper, calcium, nitrogen and phosphorus in female House Sparrows during the breeding cycle. Ardea 73:175–182

Pinowska B, Kraśnicki K (1985b) Quantity of gastroliths and magnesium and calcium contents in the body of female House Sparrows during their egg-laying period. Zesz nauk Filii UW, 48, Biol 10:125–130

Pinxten R, Eens M (1997) Invloed van verzwing en kalkgebrek op het broedsucces van de Koolmees *Parus major* op de Kalmthoutse Heide in 1995. Oriolus 63:93–96

Poulin R, Brigham RM (2001) Effects of supplemental calcium on the growth rate of an insectivorous bird, the Purple Martin (*Progne subis*). Ecoscience 8:151–156

Powlesland RG, Lloyd BD (1994) Use of supplementary feeding to induce breeding in free-living Kakapo *Strigops habroptilus* in New Zealand. Biol Conserv 69:97–106

Pulliainen E, Marjakangas A (1980) Eggshell thickness in eleven sea and shore bird species of the Bothnian Bay. Ornis Fenn 57:65–70

Pulliainen E, Kallio T, Hallaksela A-M (1978) Eating of wood by Parrot Crossbills, *Loxia pytyopsittacus*, and Redpolls, *Carduelis flammea*. Aquilo Ser Zool 18:23–27

Quesenberry KE, Hillyer EV (1994) Supportive care and emergency therapy. In: Ritchie BW, Harrison GJ, Harrison LR (eds) Avian medicine: principles and application. Wingers, Lake Worth, pp 382–416

Ramsay SL, Houston DC (1999) Do acid rain and calcium supply limit eggshell formation for Blue Tits (*Parus caeruleus*) in the U.K.? J Zool (Lond) 247:121–125

Ratcliffe DA (1970) Changes attributable to pesticides in egg breakage frequency and eggshell thickness in some British birds. J Appl Ecol 7:67–107

Raubenheimer D, Simpson SJ (2006) The challenge of supplementary feeding: can geometric analysis help save the Kakapo? Notornis 53:100–111

Raveling DG (1978) The timing of egg laying by northern geese. Auk 95:294–303

Reid BL, Weber CW (1976) Calcium availability and trace mineral composition of feed grade calcium supplements. Poult Sci 55:600–605

Remsen JV Jr, Stiles FG, Scott PE (1986) Frequency of arthropods in stomachs of tropical hummingbirds. Auk 103:436–441

Repasky RR, Blue RJ, Doerr PD (1991) Laying Red-cockaded Woodpeckers cache bone fragments. Condor 93:458–461

Reynolds SJ (1997) Uptake of ingested calcium during egg production in the Zebra Finch (*Taeniopygia guttata*). Auk 114:562–569

Reynolds SJ (1998) Uptake and disposition of calcium and water by egg-laying Zebra Finches (*Taeniopygia guttata*). D.Phil. thesis, University of Oxford

Reynolds SJ (2001) The effects of low dietary calcium during egg-laying on eggshell formation and skeletal calcium reserves in the Zebra Finch *Taeniopygia guttata*. Ibis 143:205–215

Reynolds SJ (2003) Mineral retention, medullary bone formation, and reproduction in the White-tailed Ptarmigan (*Lagopus leucurus*): a critique of Larison et al. (2001). Auk 120:224–228

Reynolds SJ, Martin GR, Cassey P (2009) Is sexual selection blurring the functional significance of eggshell coloration hypotheses? Anim Behav 78:209–215

Reynolds SJ, Mänd R, Tilgar V (2004) Calcium supplementation of breeding birds: directions for future research. Ibis 146:601–614

Richardson PRK, Mundy PJ, Plug I (1986) Bone crushing carnivores and their significance to osteodystrophy in Griffon Vulture chicks. J Zool (Lond) 210:23–43

Robbins CT (1981) Estimation of the relative protein cost of reproduction in birds. Condor 83:177–179

Robbins CT (1993) Wildlife feeding and nutrition. Academic, San Diego

Roland DA Sr (1986) Eggshell quality IV: Oystershell versus limestone and the importance of particle size or solubility of calcium source. World Poult Sci 42:166–171

Romanoff AL, Romanoff AJ (1949) The avian egg. Wiley, New York

Ryder JP (1970) A possible factor in the evolution of clutch size in Ross' Goose. Wilson Bull 82:5–13

St. Louis VL, Barlow JC (1993) The reproductive success of Tree Swallows nesting near experimentally acidified lakes in northwestern Ontario. Can J Zool 71:1090–1097

St. Louis VL, Breebaart L (1991) Calcium supplements in the diet of nestling Tree Swallows near acid sensitive lakes. Condor 93:286–294

Sadler KC (1961) Grit selectivity by the female Pheasant during egg production. J Wildl Manage 25:339–341

Sauveur B, Mongin P (1974) Effects of time limited calcium meal upon food and calcium ingestion and egg quality. Br Poult Sci 15:305–313

Savage JE (1968) Trace minerals and avian reproduction. Fed Proc 27:927–931

Scanes CG, Ottinger MA, Kenny AD, Balthazart J, Cronshaw J, Chester Jones I (eds) (1982) Aspects of avian endocrinology: practical and theoretical implications. Texas Technical University, Lubbock

Schaub M, Jenni L (2000) Fuel deposition of three passerine bird species along the migration route. Oecologia 122:306–317

Scheuhammer AM (1987) The chronic toxicity of aluminium, cadmium, mercury, and lead in birds: a review. Environ Pollut 46:263–296

Scheuhammer AM (1991) Effects of acidification on the availability of toxic metals and calcium to wild birds and mammals. Environ Pollut 71:329–375

Schifferli L (1976) Factors affecting weight and conditions in the House Sparrow particularly when breeding. D.Phil. thesis, University of Oxford

Schifferli L (1979) Das Skelettgewicht des Haussperlings *Passer domesticus* während der Eiablage und bei Jungvögeln nach dem Ausfliegen. Orn Beob 76:289–292

Schifferli L (1980) Changes in the fat reserves in female House Sparrows *Passer domesticus* during egg laying. In: Proceedings of the 17th International Ornithological Congress, Berlin, pp 1129–1135

Schmidt K-H, Zitzmann A (1990) Sprunghafter Anstieg von Brutstörungen bei Höhlenbrütern. J Ornithol 131:172–174

Scott ML, Mullenhoff PA (1970) Dietary oyster shell and egg shell quality. In: Proceedings of the Cornell Nutritional Conference for Feed Manufacturers, Ithaca, New York, pp 24–28

Scott ML, Hull SJ, Mullenhoff PA (1971) The calcium requirements of laying hens and effects of dietary oyster shell upon egg shell quality. Poult Sci 50:1055–1063

Shrubb M (1997) Historical trends in British and Irish Corn Bunting *Miliaria calandra* populations – evidence for the effects of agricultural change. In: Donald PF, Aebischer NJ (eds) The ecology and conservation of Corn Buntings *Miliaria calandra*. Joint Nature Conservation Committee, Peterborough, pp 27–41

Sibly RM, McCleery RH (1983) The distribution between feeding sites of Herring Gulls breeding at Walney Island, U.K. J Anim Ecol 52:51–68

Simkiss K (1967) Calcium in reproductive physiology. Chapman and Hall, New York

Simons PCM (1984) Major minerals in the nutrition of poultry. In: Fisher C, Boorman KN (eds) Nutrient requirements of poultry and nutritional research, Poultry Science Symposium 19, pp 141–154

Singh A, Agrawal M (2008) Acid rain and its ecological consequences. J Environ Biol 29:15–24

Six KM, Goyer RA (1970) Experimental enhancement of lead toxicity by low dietary calcium. J Lab Clin Med 76:933–942

Soares JH Jr (1984) Calcium metabolism and its control – a review. Poult Sci 63:2075–2083

Soares JH Jr (1995) Calcium bioavailability. In: Ammerman CB, Baker DH, Lewis AJ (eds) Bioavailability of nutrients for animals: Amino acids, minerals, and vitamins. Academic, Oxford, pp 95–118

Solomon SE (1997) Egg and eggshell quality. Iowa State University Press, Ames

Starck JM (1998) Structural variants and invariants in avian embryonic and postnatal development. In: Starck JM, Ricklefs RE (eds) Avian growth and development: evolution within the altricial-precocial spectrum. Oxford University Press, Oxford, pp 59–88

Stillmak SJ, Sunde ML (1971) The use of high magnesium limestone in the diet of the laying hen. Poult Sci 50:553–560

Stoddard JL, Jeffries DS, Lükeville A, Clair TA, Dillon PJ, Driscoll CT, Forsius M, Johannessen M, Kahl JS, Kellogg JH, Kemp A, Mannio J, Monteith DT, Murdoch PS, Patrick S, Rebsdorf A, Skjelkvåle BL, Stainton MP, Traaen T, van Dam H, Webster KE, Wieting J, Wilander A (1999) Regional trends in aquatic recovery from acidification in North America and Europe. Nature 401:575–578

Studier EH, Szuch EJ, Tompkins TM, Cope VW (1988) Nutritional budgets in free flying birds: Cedar Waxwings (*Bombycilla cedrorum*) feeding on Washington Hawthorn fruit (*Crataegus phaenopyrum*). Comp Biochem Physiol A 89:471–474

Sušic G (1981) Crossbill (*Loxia curvirostra* L. 1758) feeding on mortar in a wall. Larus 33–35:197–200

Symes CT, Hughes JC, Mack AL, Marsden SJ (2006) Geophagy in birds of Crater Mountain wildlife management area, Papua New Guinea. J Zool 268:87–96

Talbot CJ, Tyler C (1974) A study of the progressive deposition of shell in the shell gland of the domestic hen. Br Poult Sci 15:217–224

Taliaferro EH, Holmes RT, Blum JD (2001) Eggshell characteristics and calcium demands of a migratory songbird breeding in two New England forests. Wilson Bull 113:94–100

Taylor TG (1965) Calcium-endocrine relationships in the laying hen. Proc Nutr Soc 24:49–54

Taylor TG (1970) How an eggshell is made. Sci Am 222:88–95

Thaler E (1979) Das Aktionssystem von Winter- und Sommergoldhänchen (*Regulus regulus*, *R. ignicapillus*) und deren ethologische Differenzierung. Bonn Zool Monogr 12:1–151

Tilgar V (2002) Effect of calcium supplementation on reproductive performance of the Pied Flycatcher *Ficedula hypoleuca* and the Great Tit *Parus major*, breeding in northern temperate forests. PhD Thesis, University of Tartu

Tilgar V, Kilgas P, Mägi M, Mänd R (2008) Age-related changes in the activity of bone alkaline phosphatase and its application as a marker of prefledging maturity of nestlings in wild passerines. Auk 125:456–460

Tilgar V, Mänd R, Kilgas P, Reynolds SJ (2005) Chick development in free-living Great Tits *Parus major* in relation to calcium availability and egg composition. Physiol Biochem Zool 78:590–598

Tilgar V, Mänd R, Leivits A (1999a) Effect of calcium availability and habitat quality on reproduction in Pied Flycatcher *Ficedula hypoleuca* and Great Tit *Parus major*. J Avian Biol 30: 383–391

Tilgar V, Mänd R, Leivits A (1999b) Breeding in calcium-poor habitats: are there any extra costs? Acta Ornithol 34:215–218

Tilgar V, Mänd R, Mägi M (2002) Calcium shortage as a constraint on reproduction in Great Tits *Parus major*: a field experiment. J Avian Biol 33:407–413

Tinbergen N, Broekhuysen GJ, Feekes F, Houghton JCW, Kruuk H, Szulc E (1962) Egg shell removal by the Black-headed Gull, *Larus ridibundus* L.: a behaviour component of camouflage. Behaviour 19:74–117

Tordoff MG (2001) Calcium: taste, intake, and appetite. Physiol Rev 81:1567–1597

Turner AK (1982) Timing of laying by Swallows (*Hirundo rustica*) and Sand Martins (*Riparia riparia*). J Anim Ecol 51:29–46

Tyler C (1954) Studies on eggshells. IV. The site of deposition of radioactive calcium and phosphorus. J Sci Food Agric 5:335–339

van Breemen N, Mulder J, Driscoll CT (1983) Acidification and alkalinization of soils. Plant Soil 75:283–308

Verbeek NAM (1971) Hummingbirds feeding on sand. Condor 73:112–113

Verbeek NAM (1996) Occurrence of egg-capping in birds' nests. Auk 113:703–705

Verdoorn GH (1998) Vulture restaurants as a conservation tool. In: Boshoff AF, Anderson MD, Borello WD (eds) Vultures in the 21st century: Proceedings of a workshop on vulture research and conservation in southern Africa. Vulture Study Group, Johannesburg, pp 119–122

Vickery JA, Chamberlain DE (1999) Acid rain and birds. BTO News 223:13

Videler JJ (1995) Consequences of weight decrease on flight performance during migration. Isr J Zool 41:343–353

Wallach JD, Flieg GM (1969) Nutritional secondary hyperparathyroidism in captive birds. J Am Vet Med Assoc 155:1046–1051

Walsberg GE (1975) Digestive adaptations of *Phainopepla nitens* associated with the eating of Mistletoe berries. Condor 77:169–174

Walsberg GE (1983) Ecological energetics: What are the questions? In: Brush AH, Clark GA Jr (eds) Perspectives in ornithology: essays presented for the centennial of the American Ornithologists' Union. Cambridge University Press, Cambridge, pp 135–158

Weimer V, Schmidt K-H (1998) Untersuchungen zur Eiqualität bei der Kohlmeise (*Parus major*) in Abhängigkeit von der Bodenbeschaffenheit. J Ornithol 139:3–9

Weseloh DV, Ewins PJ, Struger J, Mineau P, Bishop CA, Postupalsky S, Ludwig JP (1995) Double-crested Cormorants of the Great Lakes: changes in population size, breeding distribution and reproductive output between 1913 and 1991. Colonial Waterbirds 18(Special issue 1):48–59

Wiley JW, Wiley BN (1979) The biology of the White-crowned Pigeon. Wildl Monogr 64:1–54

Wilkin TA, Gosler AG, Garant D, Reynolds SJ, Sheldon BC (2009) Calcium effects on life-history traits in a wild population of the great tit (*Parus major*): analysis of long-term data at several spatial scales. Oecologia 159:463–472

Wilson JE (1959) The status of the Hungarian Partridge in New York. Kingbird 9:54–57

Winkel W, Hudde H (1990) On the increasing occurrence of "Brooding on Empty Nest": findings of tits (*Parus*) and other holebreeders in various study areas in northern Germany. Vögelwarte 35:341–350

Zang H (1998) Auswirkungen des "Sauren Regens" (Waldsterben) auf eine Kohlmeisen- (*Parus major-*) Population in den Hochlagen des Harzes. J Ornithol 139:263–268

Zann R, Straw B (1984) Feeding ecology and breeding of Zebra Finches in farmland of northern Victoria. Aust Wildl Res 11:533–552

Zhang B, Coon CN (1997) The relationship of calcium intake, source, size, solubility *in vitro* and *in vivo*, and gizzard limestone retention in laying hens. Poult Sci 76:1702–1706

Chapter 3
Seasonal Metabolic Variation in Birds: Functional and Mechanistic Correlates

David L. Swanson

3.1 Introduction

The influence of seasonal changes in temperature and climate on metabolic rates in birds has been a topic of interest to ornithologists and ecophysiologists for decades (e.g., Hart 1962; Dawson 1958; Miller 1939). Because metabolic rates increase linearly with temperature in endotherms outside the thermal neutral zone, comparisons of metabolic rates among seasons or species require standardized measurements of metabolic rates. The most common of these standardized metabolic rates used for comparisons of energetics among seasons or species is basal, or resting, metabolic rate. It often serves as a baseline for comparisons of metabolic costs of activities within species, and for comparisons of the "rate of living" among species or species groups (e.g., Wiersma et al. 2007a; White et al. 2007; McKechnie et al. 2006; McKechnie and Wolf 2004; Trevelyan et al. 1990; McNab 1988; Bennett and Harvey 1987; Kersten and Piersma 1987). Theoretically, basal metabolic rate (BMR) is the minimum metabolic rate required for maintenance in endotherms. BMR is measured within the thermal neutral zone under postabsorptive digestive conditions during the resting phase of the daily cycle on resting, nongrowing, nonreproductive animals (McNab 1997). It is doubtful whether truly BMRs can ever be achieved in the laboratory, so the term resting metabolic rate (RMR) is often used to refer to such measurements, even when the standard conditions for BMR have been met. Here, I will revert to the standard terminology and consider BMR as the metabolic rate measured under the standard conditions listed above, recognizing that this may not, in fact, represent truly basal rates (Table 3.1).

D.L. Swanson (✉)
Department of Biology, University of South Dakota, 414 East Clark Street, Vermillion, SD 57069, USA
e-mail: david.swanson@usd.edu

C.F. Thompson (ed.), *Current Ornithology Volume 17*, Current Ornithology,
DOI 10.1007/978-1-4419-6421-2_3, © Springer Science+Business Media, LLC 2010

Table 3.1 Definitions of commonly used terms

Term	Definition
Basal metabolic rate (BMR)	Minimum metabolic rate required for maintenance; measured as the metabolic rate at thermoneutrality in resting, postabsorptive, nongrowing birds in the resting phase of the daily cycle
Summit metabolic rate (M_{sum})	Maximum metabolic rate achieved during cold exposure
Maximum metabolic rate (MMR)	Maximum metabolic rate achieved during any form of activity. Maximal metabolic rates during locomotion are generally higher than those during cold exposure in birds
Field metabolic rate (FMR)	Metabolic rate for free-living birds engaged in normal daily activities. When expressed in terms of energy use per day, FMR is the Daily Energy Expenditure (DEE)
Cold tolerance	The time period over which a bird can maintain T_b by thermogenesis (principally shivering) at a given level of cold exposure
Shivering endurance	The time period over which a bird can continue shivering thermogenesis at a given level of cold exposure, essentially equivalent to cold tolerance
Temperature at cold limit (T_{CL})	The temperature producing hypothermia during exposure of an individual bird to a declining series of temperatures
Acclimation	Exposure of birds to controlled climatic conditions *in the laboratory*
Acclimatization	Exposure of birds to natural climatic conditions *in the field*

Another standardized measure of metabolic rate is the maximal metabolic rate during activity (MMR). Comparisons of MMR in birds are not currently possible, as MMR during flight has not been convincingly measured for any bird because of technical difficulties associated with verifying that oxygen consumption during sustained flight has actually reached maximum values. However, recent studies employing hovering flight (Chai and Dudley 1995, 1996) or hop-flutter wheels (Wiersma et al. 2007b; Pierce et al. 2005; Chappell et al. 1999) to induce maximum metabolic rates appear potentially promising in this regard, although such measurements are probably not practical for all volant birds. Thus, I will not examine maximal activity-induced metabolic rates in birds in this review. However, a similar standardized metabolic measure that has become more common in recent seasonal comparisons is summit metabolism (M_{sum}) or maximal thermogenic capacity. Here I define M_{sum} as the maximum metabolic rate attained by birds under cold exposure. Historically, summit metabolism measurements were rarely undertaken because of the very cold temperatures required to elicit M_{sum} (but see Saarela et al. 1995; Dawson and Carey 1976; Hart 1962). More recently, helium–oxygen atmospheres (helox) that facilitate heat loss without impairing metabolic function have been used to generate M_{sum} measurements (Rosenmann and Morrison 1974). The utility of M_{sum} measurements emerge not only from the indication that they provide for the maximal capacity for thermogenesis in the cold, but also from the general positive association between M_{sum} and endurance at submaximal levels of cold exposure (Marsh and Dawson 1989). In addition, M_{sum} and MMR are significantly and positively correlated in some mammals (Hayes and Chappell 1990), so they may both serve as effective general measures of aerobic capacity in endotherms, although M_{sum} and MMR (measured by hop-flutter wheel) were not correlated in tropical birds (Wiersma et al. 2007b).

Most research attention relating to seasonal variation in metabolic rates in birds has been paid to metabolic variation between summer and winter in strongly seasonal climates, as an aspect of seasonal acclimatization to cold and its attendant thermogenic demands. However, the precise nature of the association between variation in metabolic rates and variation in cold tolerance in birds remains uncertain (reviews by Dawson and O'Connor 1996; Dawson and Marsh 1989; Marsh and Dawson 1989). Winter increases in M_{sum} and/or BMR are correlated with improved cold resistance in birds and may be adaptive in this context (O'Connor 1995a; Swanson 1990a). In addition, M_{sum} and cold tolerance are correlated in birds on both intraspecific and interspecific bases (Swanson and Liknes 2006; Swanson 2001).

Surprisingly little effort has been directed toward measurement of metabolic rates of birds in migratory disposition. A substantial amount of effort has been devoted to defining the physiological and biochemical adjustments associated with migration (McWilliams et al. 2004; Gwinner 1990; Dawson et al. 1983a), but this effort has been primarily directed at describing how flight muscles might increase power and endurance during migration and such changes have not been previously tied to changes in organismal metabolism. Recent work, however, has begun to examine potential seasonal variation in basal and summit metabolic rates of migrating birds (Vezina et al. 2007; Battley et al. 2000; Swanson and Dean 1999; Piersma et al. 1995).

Another measure of metabolic rate in free-living birds that has been used for seasonal comparisons is the field metabolic rate (FMR). FMR describes the energetic costs of normal daily activities in birds under natural environmental conditions. FMR is usually measured by two methods, doubly labeled water and time–energy budgets. The theory behind these two methods has been well developed and both methods have been used to measure energy use during various periods of the annual cycle in birds (Speakman 1997; Weathers and Sullivan 1993). The doubly labeled water method measures carbon dioxide production, so it provides a more direct measure of FMR than the time-activity budget method, which sums the metabolic costs of different activities observed in free-living birds. Nevertheless, if metabolic costs of different activities are carefully measured in the lab and microclimates are carefully measured in the field, the time–energy budget and doubly labeled water methods are usually in relatively close agreement (Webster and Weathers 2000; Weathers and Sullivan 1993; Goldstein 1990).

This paper reviews the recent literature on seasonal variation in basal and summit metabolic rates associated with seasonal acclimatization and migratory disposition in birds. In addition, seasonal variation in FMRs will be reviewed in the context of how seasonal changes in energy expenditure, and in the partitioning of energy expenditure, might impact BMR and M_{sum} (or MMR). I will also discuss the potential functional significance of this seasonal variation to meeting the energetic demands of migration and thermogenesis in cold climates. Next, I will discuss the potential mechanisms underlying seasonal variation in BMR and M_{sum}. I will also try to illuminate some potential topics or avenues for future research. The overall goal is to integrate biochemical and physiological adjustments at the cellular and tissue levels, organismal physiological variation, and ecology.

3.2 The Variable Thermal Environment Encountered by Birds

3.2.1 Climate and Thermostatic Costs

Temperature is one of the major modifiers of metabolic level in endothermic animals, so it is necessary to examine the climatic conditions to which birds are exposed throughout their annual cycle. Birds nesting in temperate-zone or arctic latitudes, where climatic conditions deteriorate and productivity is reduced in winter, respond to seasonally changing environments by two major strategies, permanent residency and acclimatization or migration to more favorable climates. Many small mammals exhibit a third strategy, winter dormancy or hibernation, but such a strategy has been claimed for only one bird, the Common Poorwill (*Phalaenoptilus nuttalii*). These birds become inactive and may remain in one location for up to 3 months (Brigham et al. 2006; French 1993; Jaeger 1949). In the torpid condition, skin temperature may drop as low as 2.8°C (Brigham 1992), although body temperature may fluctuate markedly on a daily basis (e.g., up to 11°C per day, French 1993). Despite the energetic savings attributable to such a hibernation strategy, it has not been documented for any other bird species. However, torpor and regulated hypothermia are used by a wide variety of small birds on a daily basis and this can also result in substantial energetic benefit (see McKechnie and Lovegrove 2002, 2006; Reinertsen 1996, for reviews).

Species that are resident in temperate-zone or arctic latitudes, or those that migrate only relatively short distances so that they still winter in cold climates, are faced with marked seasonal changes in cold exposure and thermostatic costs (e.g., Cooper 2000; Wiersma and Piersma 1994). Winter climates in which these birds reside can be severe. Moreover, environmental productivity is reduced in these regions in winter compared to summer and daylength is short, so species employing this strategy are faced with the interacting effects of short days available for foraging, long nights of forced fasting, relatively low food availability, and cold temperatures (Marsh and Dawson 1989). Such factors may combine to make the winter environment thermally stressful for resident birds. If seasonal changes in temperature or climate are a major factor driving seasonal adjustments of physiology, then species wintering in cold climates should demonstrate substantial seasonal physiological adjustment.

Migratory species, particularly those that winter in tropical to subtropical climates, move to more favorable climates in winter and thereby reduce their exposure to the climatic deterioration and declining food base associated with wintering on or near the breeding grounds. This, in turn, reduces winter thermoregulatory costs relative to those for birds wintering in colder climates (Wiersma and Piersma 1994). As might be expected, based on the severity of winter climate, the percentages of birds migrating varies with latitude such that higher latitudes show an increased proportion of birds moving out of the area in winter. For example, in North America the percentage of birds migrating south in winter varies from 12% in Florida (25°N) to 87% at Ellesmere Island (80°N) (Newton and Dale 1996). Thus, migrating birds escape the potentially stressful environmental conditions present on their breeding grounds during the winter.

During migration, however, birds may encounter adverse weather conditions, although they are still not exposed to the seasonal extremes of temperature that cold climate-wintering species encounter. For birds wintering in tropical or subtropical climates, which migrate later in spring and earlier in fall than shorter-distance migrants (Hagan et al. 1991), average temperatures during spring migration are often lower than those during fall migration (O'Reilly and Wingfield 1995). This seasonal difference is moderated for shorter-distance migrants. For example, mean temperatures surrounding median spring passage dates of the Neotropical migrants Warbling Vireo *Vireo gilvus* and Yellow Warbler *Dendroica petechia* in southeastern South Dakota were 5–6°C colder than temperatures surrounding median fall passage dates (Swanson and Dean 1999). For mild temperate-zone migrants Ruby-crowned Kinglet *Regulus calendula* and Yellow-rumped Warbler *Dendroica coronata,* migrating through the same area, mean temperatures around spring and fall passage dates differed by only 1–2°C.

Because of the differences in seasonal temperature extremes encountered by cold climate residents and migrants, if temperature drives seasonal changes in metabolic rates in birds, then migrants should show less metabolic variation than residents. However, migration is also an energetically expensive endeavor and physiological adjustments promoting endurance muscular activity during migration occur in birds and might influence organismal metabolic rates (Vezina et al. 2007; Dawson et al. 1983a). Knowledge of how such migratory adjustments relate to seasonal variation in organismal metabolic rates is rather rudimentary at present (Vezina et al. 2007; Battley et al. 2000; Swanson and Dean 1999; Piersma et al. 1995).

3.2.2 Seasonal Variation in Field Metabolic Rates

For birds wintering in cold climates, higher thermostatic costs in winter might be expected to elevate overall energy demands in winter relative to other seasons. However, it is often assumed that energy expenditure of parent birds should peak during the breeding season, in support of breeding and parental care costs, to maximize the number of offspring produced and thereby maximize fitness. According to this scenario, FMR (or daily energy expenditures, DEE) should be greatest during the breeding season, or at least during the most expensive stage of the nesting cycle. Two hypotheses have been proposed to describe seasonal patterns of energy expenditure by birds, the increased demand hypothesis and the reallocation hypothesis (Masman et al. 1986). The increased demand hypothesis posits that breeding and parental care activities constitute a higher energetic demand than that at other times of the year, such that FMR during the breeding season, or at least during the most demanding period of the breeding season, is the highest of the annual cycle. The reallocation hypothesis contends that the moderate temperatures and abundant food base during the breeding season reduce thermostatic and foraging costs so that energy used for these purposes can be reallocated to support costs of breeding and parental care. According to this hypothesis, summer and winter FMR should be similar. However, for small birds wintering in cold climates the high costs of thermoregulation over prolonged periods might be expected to elevate FMR in winter relative to other seasons.

Weathers and Sullivan (1993) and Dawson and O'Connor (1996) reviewed seasonal studies of FMR in birds and concluded, for birds studied to that time (eight species total), that FMR during the breeding season was as high or higher than that at other times of the year, thus supporting the increased demand or reallocation hypotheses. Also in support of this conclusion, Anava et al. (2000) documented approximately equal FMR in summer and winter for nonbreeding Arabian Babblers *Turdoides squamiceps* in the northern Arabian Saharan desert in Israel. For breeding Arabian Babblers, FMR was higher in birds during the spring breeding season than nonbreeding birds in summer and winter, and this difference resulted from higher daytime energy expenditure; nocturnal energy expenditure did not vary among breeding, summer, and winter (Anava et al. 2002). Williams (2001) also found stable FMR over the annual cycle in Dune Larks (*Mirafra erythroclamys*) from the Namib desert in southern Africa. However, several recent studies of small birds have documented higher FMR in winter than in summer. Nonbreeding Verdins *Auriparus flaviceps* in southeastern Arizona had 47% higher FMR in winter than in summer, but costs of breeding were not incurred by summer birds in this study (Webster and Weathers 2000). FMR of White-crowned Sparrows *Zonotrichia leucophrys* on the central California coast was 17% higher in winter than during incubation or nestling periods in summer, despite the moderate winter climate at this location (Weathers et al. 1999). In addition, Mountain Chickadees *Poecile gambeli* and Juniper Titmice *Baeolophus griseus* in northern Utah showed 36 and 104% increases in FMR, respectively, in winter relative to summer (Cooper 2000). Carolina Chickadees *Poecile carolinensis* from Ohio also showed elevated FMR in winter relative to summer, with winter values 42% greater than summer values (Doherty et al. 2001). Thus, winter FMR in some birds may be higher than breeding season FMR and these results conflict with those from earlier studies.

Are there any patterns that emerge from these studies on seasonal variation in FMR in birds? Cooper (2000) noted that previous studies of seasonal variation in FMR had largely been conducted on birds wintering in relatively mild winter climates, where winter increases in thermostatic costs would be moderate. In addition, of the species in the Weathers and Sullivan (1993) and Dawson and O'Connor (1996) reviews, those wintering in relatively cold climates had body masses above 160 g. Because large body size results in low surface area to volume ratios and low thermal conductance, thermostatic costs should vary much less seasonally in these birds than in small birds. Thus, the emerging pattern of seasonal variation in FMR appears to be seasonal stability or higher FMR during the breeding season than during winter for large birds or for birds wintering in mild winter climates, but higher FMR in winter than during the breeding season for small birds wintering in cold climates. This emphasizes the prominent role of thermoregulatory costs in the winter energy budget of small birds in cold climates and underscores the importance of the metabolic adjustments by which small birds meet these demands. These adjustments, in turn, are likely to influence BMR and M_{sum}.

3.3 Patterns and Functional Significance of Seasonal Metabolic Variation

3.3.1 Thermogenic Mechanisms

Before discussing patterns of seasonal variation in metabolic rates, a brief review of the mechanisms of heat production in birds is in order because increased thermoregulatory demands have been implicated as a factor driving winter increases in metabolic rates (Dawson and O'Connor 1996). Heat is generated in resting birds primarily, if not exclusively, by shivering (Marsh and Dawson 1989). The flight muscles, pectoralis and supracoracoideus, comprise the largest muscle group in the body of most birds, and range in size from about 15 to 25% of total body mass (Hartman 1961). Flight muscles thus contribute greatly to shivering thermogenesis and, as a consequence, have been the most extensively studied muscle group regarding shivering thermogenesis. However, in some species leg muscles may also contribute importantly to shivering thermogenesis (Marjoniemi and Hohtola 1999; Carey et al. 1989). Birds shiver isometrically so that antagonistic muscle groups work against each other (Hohtola 1982). This has the decided advantage of creating little disruption of the insulative layer at the body surface. Generally, increases in electrical activity in shivering muscles are linearly related to increases in metabolic rate during shivering and EMG activity and oxygen consumption both increase as ambient temperature declines (e.g., Hohtola et al. 1998; Tøien 1992).

Nonshivering thermogenesis (NST) has been claimed for several bird species, based on discrepancies between oxygen consumption and EMG activity (see Reinertsen 1996; Dawson and O'Connor 1996; Duchamp et al. 1993; for reviews). The occurrence of NST in birds, however, is controversial (Hohtola 2002; Marsh 1993) and if it does occur, the site of action is unknown, although skeletal muscle appears to be a likely candidate (Reinertsen 1996). Birds lack brown fat, the principal site of NST in mammals, and they also lack the uncoupling protein (UCP 1) responsible for dissociation of electron transport from ATP production in the mitochondrion that allows the thermogenic function of brown fat in mammals (Mezentseva et al. 2008; Emre et al. 2007; Brigham and Trayhurn 1994; Saarela et al. 1989a, 1991; Olson et al. 1988). Thus, if NST occurs in birds, its primary site is different from that in mammals. Birds do possess an UCP gene (*avUCP*) that is highly homologous to mammalian *UCP 2* and *UCP 3* and is expressed in skeletal muscle, heart and liver (Mozo et al. 2005; Raimbault et al. 2001; Vianna et al. 2001). Expression of *avUCP* increases in cold-acclimated and glucagon-treated Muscovy ducklings (*Cairina muschata*) (Raimbault et al. 2001). Cold exposure in Swallow-tailed Hummingbirds (*Eupetomena macroura*) elicited torpor and *avUCP* expression increased in torpid birds (Vianna et al. 2001). Moreover, cold exposure in broiler chicks increased plasma triiodothyronine (T_3) levels and *avUCP* expression (Collin et al. 2003a) and treatment of broiler chicks with T_3 increased, whereas treatment with thyroid hormone decreased, *avUCP* expression (Collin et al. 2003b). These data suggest some role for avUCP in energy expenditure in birds, mediated by T_3, but Mozo et al. (2005) suggest that the primary role for avUCP may the control of production of reactive oxygen species by mitochondria, rather than thermogenesis.

Maximum metabolic rates during cold exposure in birds generally range from about 3 to 8-times BMR (Dutenhoffer and Swanson 1996; Saarela et al. 1995; Marsh and Dawson 1989). The largest factorial increment in metabolic rate recorded during cold exposure in birds is from summer acclimatized House Sparrows (*Passer domesticus*) from Wisconsin, USA, where M_{sum} exceeds BMR by ninefold (Arens and Cooper 2005a). Interestingly, winter acclimatized House Sparrows from this same population exhibited a factorial increment of only 6.9. The second largest factorial increment in the cold (8.4) is from a South American hummingbird, the Green-backed Firecrown (*Sephanoides sephanoides*; López-Calleja and Bozinovic 1995). Metabolic expansibility (M_{sum}/BMR) during cold exposure is usually less than that during locomotion (MMR/BMR), where metabolic expansibilities generally range from about 8 to 14-times BMR for running or flying birds (Brackenbury 1984). Klaassen et al. (2000) suggest that long-term steady-state flights may be somewhat cheaper, as metabolic rates during extended flights in a Thrush Nightingale *Luscinia luscinia* exceeded BMR by only 5.5-times. These values, however, do not represent maximal aerobic metabolic rates during locomotion, which can be much higher. For example, Bundle et al. (1999) recorded a metabolic expansibility for Rheas *Rhea americana* running on an inclined treadmill of 36-times BMR. Flight metabolic expansibility values exceeding 20 have been claimed for some birds, including migrating grebes *Podiceps nigricollis* (based on fat depletion rates; Jehl 1994), pigeons *Columba livia* carrying loads (Gessaman et al. 1991; Gessaman and Nagy 1988), and some small passerines during short flights (Nudds and Bryant 2000; Tatner and Bryant 1986).

The full aerobic capacity of the flight muscles is thus evidently not available for shivering thermogenesis. Marsh and Dawson (1989) suggest three possible reasons for the difference between M_{sum} and maximum exercise-induced metabolic rate in birds. First, the muscle mass recruited may be less for shivering than for flight if mechanical constraints on force production during shivering exist. Because birds shiver isometrically, force production by the smaller of the antagonistic muscle groups (supracoracoideus in the case of flight muscles) may limit force production by the larger muscle group (pectoralis in the case of flight muscles). The high metabolic expansibility during cold exposure for the only hummingbird for which M_{sum} has been measured, *Sephanoides sephanoides* (López-Calleja and Bozinovic 1995), is noteworthy in this regard because hummingbirds have the largest supracoracoideus mass relative to body mass among birds because of their hovering flight style (Mathieu-Costello et al. 1992). Second, because body temperature of birds is often slightly reduced during shivering at very cold temperatures and slightly elevated during flight, temperature effects on metabolic processes could result in elevated flight metabolic rates relative to shivering metabolic rates. Finally, the nearly constant contraction in shivering muscles may restrict blood flow to the muscle during shivering relative to blood flow to muscles during flight where the muscles intermittently contract and relax. Such a restriction in blood flow could reduce oxygen supply to shivering muscles and thereby limit maximum aerobic metabolic rates during shivering.

Heat produced as a by-product of digestion (heat increment of feeding) or activity may substitute for thermoregulatory heat production in some birds (Dawson and O'Connor 1996). Partial substitution of activity- or digestion-generated heat for thermostatic costs occurs when metabolic rates at temperatures below thermoneutrality are elevated above those for activity or digestion alone, but not to the combined level of activity or digestion costs plus thermostatic costs for resting birds. Complete substitution occurs when metabolic rates for activity or digestion at a given temperature below thermoneutrality are indistinguishable from those for thermoregulation at rest. Partial or complete substitution of activity-induced heat production for thermostatic costs has been documented for White-crowned Sparrows *Zonotrichia leucophrys* (Paladino and King 1984), Verdins *Auriparus flaviceps* (Webster and Weathers 1990), Gambel's Quail *Callipepla gambelii* (Zerba and Walsberg 1992), Red Knots *Calidris canutus* (Bruinzeel and Piersma 1998), Ruby-throated Hummingbirds *Archilochus colubris* (Chai et al. 1998), House Finches *Carpodacus mexicanus* (Zerba et al. 1999), and Black-capped Chickadees *Poecile atricapillus* (Cooper and Sonsthagen 2007). Partial to complete substitution of the heat increment of feeding for thermostatic costs has also been reported for Eurasian Kestrels *Falco tinnunculus* (Masman et al. 1989), House Wren chicks *Troglodytes aedon* (Chappell et al. 1997), Thick-billed Murres *Uria lomvia* (Hawkins et al. 1997), pigeons (Rashotte et al. 1999), and Tawny Owls *Strix aluco* (Bech and Præstang 2004). The net result of this substitution is that overall thermostatic costs are reduced in active or digesting birds. However, Kaseloo and Lovvorn (2003) found that Mallards (*Anas platyrhynchos*) eating grain voluntarily in small meal sizes showed no substitution of the heat increment of feeding for thermoregulation.

3.3.2 BMR and Seasonal Acclimatization

The positive relationship between latitude and BMR in terrestrial (Broggi et al. 2007; Weathers 1979) and marine (Ellis 1984) birds suggests that differences in BMR among birds are related to climatic variation. In addition, tropical resident species often have lower BMR than temperate-zone species (Wiersma et al. 2007a; Hails 1983), although some of this variation may be related to habitat, as tropical species from shaded forest areas tend to have average BMR while those from open sunny habitats have reduced BMR (Merola-Zwartjes 1998; Weathers 1997; Vleck and Vleck 1979). In addition, for species that migrate to tropical regions or for those that occupy geographic ranges encompassing both temperate-zone and tropical populations, BMR tends to be lower in the tropics (Kersten et al. 1998; Lindström 1997; Klaassen 1995; Kersten and Piersma 1987).

Laboratory acclimation experiments have shown that temperature can influence BMR, with cold acclimation increasing BMR and warm acclimation decreasing BMR in certain captive species (see McKechnie 2008, for review; also see Gelineo 1964). Although BMR in winter is usually higher than in summer for wild birds in seasonal climates, it may also be lower or seasonally stable (see McKechnie 2008; Dawson and O'Connor 1996, for reviews), so winter increases in BMR are not required for seasonal acclimatization to cold. Weathers and Caccamise (1978) suggested that seasonal variation in BMR is related to body mass, with large birds (>200 g) generally showing winter decreases in BMR, while small birds generally show winter increases in BMR (Fig. 3.1). However, McKechnie (2008) critically reevaluated Weathers and Caccamise's (1978) data and included additional data meeting strict requirements for BMR and found no evidence of a relationship between body mass and the magnitude of seasonal variation in BMR for birds, although he noted that sample sizes for species >30 g were too low at present for definitive conclusions.

How much does BMR vary seasonally or with temperature acclimation in those species showing labile BMR? Gelineo (1964) reported changes in BMR in captive birds ranging from 10 to 85% upon cold acclimation in the laboratory. Similarly, more recent data show cold acclimation-induced increases in BMR ranging from 5 to 42% (McKechnie 2008). Interestingly, BMR of Rufous-collared Sparrows (*Zonotrichia capensis*) showed a greater response to cold acclimation in populations from seasonally variable environments than in those from seasonally stable environments, suggesting that seasonal phenotypic flexibility in BMR is associated with climatic variability (Cavieres and Sabat 2008). The maximum degree of seasonal variation in BMR for wild birds to date was documented for House Sparrows, where BMR was 64% higher in winter than in summer (Arens and Cooper 2005a). According to the allometric equation of Weathers and Caccamise (1978), winter BMR/summer BMR ratios of 1.24, 1.16, and 1.05 would be expected for birds of 10, 25, and 100 g body mass, respectively.

Because winter elevations of BMR in small birds are often correlated with improvements in cold tolerance (e.g., Liknes et al. 2002; Liknes and Swanson 1996;

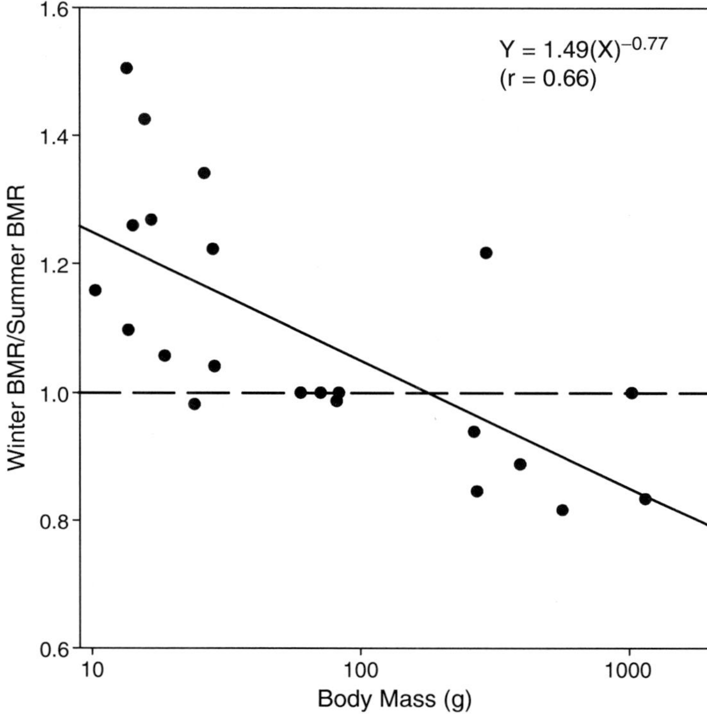

Fig. 3.1 Relationship between body mass and winter/summer BMR ratio in birds. The *dashed line* represents equality between summer and winter BMR. Birds less than about 200 g tend to show elevated BMR in winter relative to summer, while the opposite is generally true for larger birds. Redrawn from Weathers and Caccamise (1978), with permission

Cooper and Swanson 1994; Swanson 1991a; Weathers and Caccamise 1978), this suggests that changes in BMR are functionally correlated with changes in cold tolerance. Dawson and O'Connor (1996) suggested that an elevation of BMR in winter appears unnecessary and energetically expensive for cold defense given the substantial capacity of birds for regulatory thermogenesis. They mentioned two possibilities by which winter increases in BMR could benefit cold tolerance in birds directly. First, winter increases in BMR could serve as an emergency response for protection of peripheral tissues from cold injury. Second, an elevation of BMR could lower the threshold temperature for initiation of shivering, thus decreasing energetic costs of thermoregulation. Neither of these possibilities seem particularly likely, as shivering thermogenesis should provide sufficient protection of peripheral tissues from cold injury and an elevated BMR entails an energetic cost that offsets the benefit from a decreased threshold for shivering thermogenesis. Regarding the relationship of increases in BMR with increases in cold tolerance, Dawson and O'Connor (1996) suggest that it is currently impossible to distinguish whether increases in BMR in winter contribute to increases in cold tolerance, are a by-product of them, or are a separate response. The by-product possibility seems most likely, as the increased metabolic machinery required for enhanced thermogenesis in winter probably entails higher maintenance costs (Swanson 1991a). Such a relationship should be manifested by a positive correlation between BMR and M_{sum} in birds. Some studies have found that BMR and M_{sum} are significantly correlated in birds on an interspecific basis (Rezende et al. 2002; Dutenhoffer and Swanson 1996), but Wiersma et al. (2007b) found no significant interspecific correlation between M_{sum} and BMR for tropical birds. To my knowledge, the relationship between BMR and M_{sum} on an intraspecific basis is unstudied in birds. Potentially pertinent to this question, however, Chappell et al. (1999) found that BMR and maximal exercise-induced metabolic rates were significantly and positively correlated in juvenile House Sparrows, but not in adults.

Most treatments of seasonal variation in BMR in birds have focused on its association with winter acclimatization, but variation in BMR related to the energetics of reproduction might also be expected. In this regard, decrements of BMR in the nonmigratory season or with summer acclimatization are common in birds (Dawson and O'Connor 1996). Such decreases in BMR during periods of the year where energy demands for migration or thermoregulation are reduced may be directly adaptive, functioning to decrease maintenance costs of metabolically active tissues during these periods. Ambrose and Bradshaw (1988) found that White-browed Scrubwrens (*Sericornis frontalis*) from arid environments exhibited a BMR that was 19% lower in summer than in winter, while scrubwrens from more mesic environments did not undergo seasonal variation in BMR. In addition, birds from the arid environments had lower BMR than birds from the more mesic environments in summer, but not in winter. These data suggest that the decrease in summer BMR for scrubwrens from arid environments could be directly adaptive, allowing reductions in energy expenditure, heat production, and respiratory evaporative water loss in hot, dry environments (Ambrose and Bradshaw 1988).

Another physiological adjustment during the breeding season that could contribute to decreases in BMR is mass loss during reproduction. This is a common occurrence among birds, although such mass losses are often restricted to females that are largely responsible for chick brooding and provisioning (Moreno 1989). Two alternative hypotheses have been proposed to explain reproductive mass losses in birds, the cost of reproduction hypothesis and the adaptive mass loss (or programmed anorexia) hypothesis (e.g., Golet and Irons 1999). The cost of reproduction hypothesis contends that the heavy workload associated with reproduction, incubation, brooding, and chick provisioning results in nutritional stress and decreases energetic condition, particularly in female birds. Thus, the decline in body condition and fat stores is viewed as a cost to reproduction. According to this hypothesis, mass loss should be greatest during the period of heaviest workload (i.e., feeding nestlings). Patterns of mass loss consistent with this hypothesis have been documented for a number of bird species (e.g., Nagy et al. 2007; Williams et al. 2007; Moe et al. 2002; Holt et al. 2002; Chastel and Kersten 2002; Murphy et al. 2000; Golet and Irons 1999; Bryant 1988; Nur 1984). Alternatively, mass loss during breeding could be adaptive if it reduces energetic costs (BMR and flight costs) of adults involved in chick rearing. According to this hypothesis, mass loss should not be coincident with the period of highest workload, but should occur prior to or at the initiation of chick provisioning so that flight costs (and BMR) are lower during this period. A number of studies have documented this pattern of mass loss in breeding birds (e.g., Blem and Blem 2006; Kullberg et al. 2002; Cichon 2001; Phillips and Furness 1997; Sanz and Moreno 1995; Croll et al. 1991; Ricklefs and Hussell 1984; Freed 1981; Norberg 1981).

Variation of BMR in birds has been correlated with variation in lean mass or with variations in organ masses in several studies (e.g., Liu and Li 2006; Vezina and Williams 2003, 2005; Bech and Ostnes 1999; Chappell et al. 1999; Piersma et al. 1996a; Scott et al. 1996; Daan et al. 1990). Furthermore, a number of studies have estimated the reductions in BMR and flight costs potentially attributable to repro-ductive mass losses. These reductions range from about 5 to 25% for BMR and from about 4 to 10% for flight costs (Phillips and Furness 1997; Jones 1994; Croll et al. 1991). However, these estimates were generated from allometric equations predicting metabolic rates as a function of body mass and changes in body compo-sition associated with reproductive mass losses were not quantified. Thus, these indirect estimates are of little value for describing the energetic consequences of breeding season mass losses. If mass losses were primarily from metabolically inactive tissues (e.g., fat), the reductions in BMR may be overestimated. However, loss of muscle protein during the egg-laying period has been documented for a number of birds, although its occurrence is not universal (Cottam et al. 2002; Houston et al. 1995; Jones 1991), and muscle protein levels depleted during the laying period may remain low during incubation (Mawhinney et al. 1999). Regardless of how breeding season mass losses are accomplished, significant ener-getic savings could accrue from decreases in flight costs and BMR during the reproductive season (Norberg 1981).

Few studies have directly measured variation in RMR over the breeding season in birds to evaluate how such changes might associate with variation in body mass. Nilsson and Råberg (2001) found that RMR in Great Tits (*Parus major*) increased, relative to winter levels, by 12% during nest building, 27% during egg production, and 20% during chick provisioning stages of the nesting cycle. RMR also varied among nonbreeding, egg-laying, and chick provisioning periods in European Starlings (*Sturnus vulgaris*), but relative RMR among the stages differed among years, with each stage being highest during one of 3 years of measurement (Vezina and Williams 2002). Black-legged Kittiwakes (*Rissa tridactyla*) exhibited a 26% reduction in RMR during the chick provisioning stage compared to pre-breeding and incubation stages of the nesting cycle, and this was associated with a concomi-tant increase in FMR and decrease in body mass (Fyhn et al. 2001). The body mass decrease between incubation and chick provisioning in kittiwakes is accompanied by reductions in plasma triiodothyronine levels and kidney metabolic intensity (Rønning et al. 2008), suggesting that a downregulation of metabolic intensity, in addition to decreased body mass, may contribute to the reduced BMR during chick rearing. However, Vezina and Williams (2005) found that mass-specific metabolic intensity (measured by citrate synthase activity) did not vary consistently with mass changes in several organs across the breeding season in starlings, so mass and metabolic intensity of organs do not necessarily vary in tandem. This result is also supported by the finding that Red Knots (*Calidris canutus*) show a reduced BMR, despite increases in body mass, lean mass and gizzard mass when switched to a lower quality diet (Piersma et al. 2004).

3.3.3 BMR and Migratory Disposition

Migration involves sustained bouts of an energetically expensive behavior (flight) that are not immediately balanced by feeding and resting. Thus, migration results in physiological changes promoting endurance flight in birds (McWilliams et al. 2004; Gwinner 1990; Dawson et al. 1983a). Organismal metabolic rates might be expected to increase with migratory disposition because of the physiological adjustments promoting endurance flight. Several studies have examined seasonal variation in BMR (on a whole-organism basis) associated with migration (Table 3.2). Piersma et al. (1995) found that arctic-breeding Red Knots *Calidris canutus* held as outdoor-captives on the western European wintering grounds, demonstrated seasonal variation in BMR (up to 2.1-fold), with values in late spring and early summer (spring migration and initiation of breeding) higher than the annual minimum value during winter (Fig. 3.2). Similarly, Weber and Piersma (1996) showed that BMR in outdoor captive knots decreased markedly, along with mass, over a 31-day period following the attainment of peak mass associated with the spring migration period. Lindström (1997) found that BMR was higher in populations of autumn migrants than in populations on the African wintering grounds for three species of shorebirds, although the Sanderling (*Calidris alpina*) showed no seasonal

Table 3.2 Variation in BMR associated with migratory disposition in birds

| | Mass (g) | | | BMR (mL $O_2 \cdot min^{-1}$) | | | | |
| | Annual | | | Annual | | | | |
Species	Spring	Low	Fall	Spring	Low	Fall	H/L	References
Shorebirds								
Calidris canutus	165	115[a]	115	3.97	2.06[a]	2.24	2.1	Piersma et al. 1995
C. alpina	–	35	44	–	1.04	1.58	1.51	Lindström 1997
C. ferruginea	–	52	51	–	1.64	1.85	1.27	Lindström 1997
C. alba	–	47	47	–	1.58	1.58	1.0	Lindström 1997
Arenaria interpres	–	100	87	–	1.67	2.75	1.64	Lindström 1997
Passerines								
Dendroica coronata	13	–	11	0.84	–	0.64	1.31	Swanson and Dean 1999

The Annual Low column refers to the time period providing the lowest BMR over the annual cycle. For the shorebirds above, which are arctic-breeders and tropical-winterers, this period is in winter
[a]Values estimated as averages for three individuals from Fig. 3 in Piersma et al. 1995
H/L: the annual high BMR divided by the annual low BMR

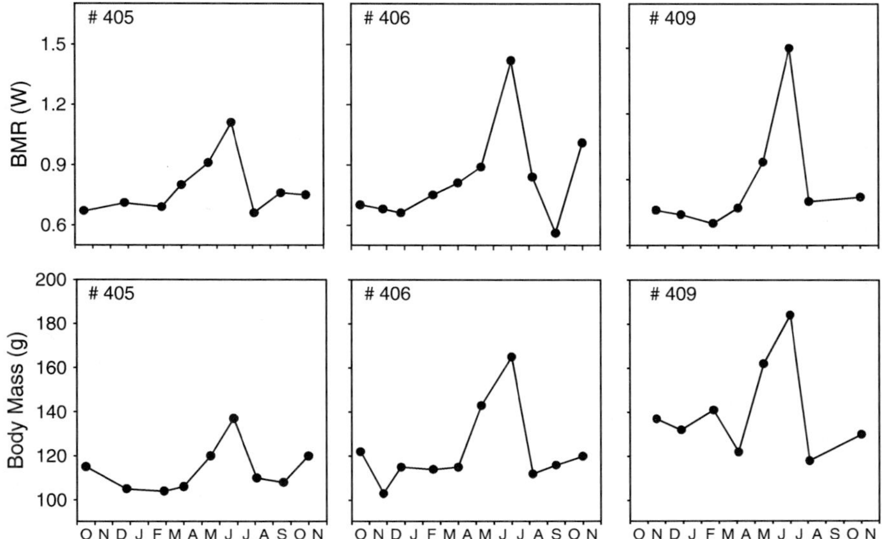

Fig. 3.2 Annual variation in body mass and BMR (in Watts) for three individual captive Red Knots, *Calidris canutus*. Both mass and BMR increased during the spring migration period and peaked during late June/early July, which coincides with the early breeding season on their arctic breeding grounds. Redrawn from Piersma et al. (1995), with permission

variation in BMR. Redshanks *Tringa totanus* also demonstrated elevated BMR in association with migratory fattening (Scott et al. 1996). Recently, Vezina et al. (2007) demonstrated that BMR in Red Knots varied over the cycle of fattening and mass loss surrounding the spring migration period and that changes in BMR were positively associated with changes in body mass and pectoral muscle mass. In addition to the shorebird data, BMR in Yellow-rumped Warblers increased by 31% in spring relative to fall migration (Swanson and Dean 1999).

The general pattern emerging from these studies is that whole-organism BMR does increase with migratory disposition in birds, at least during spring migration, and that elevated BMR accompanies the increment in body mass during migration (Fig. 3.2). Because only a few species have been studied, these conclusions regarding seasonal variation in BMR with migratory disposition must remain tentative at present. Further research, including BMR measurements from a wider variety of taxa, is needed to determine if these general patterns are robust. Additional specific research questions in need of study include: (1) Is the migratory elevation of BMR detected in shorebirds common to other taxa as well? (2) How does BMR compare between spring and fall migration? (3) Does variation of BMR with migration differ between short-distance and long-distance migrants, perhaps as a result of differential storage and use of protein and fat prior to and during migratory flights (Piersma and Jukema 2002; Klaassen et al. 1997; Klaassen 1996; Klaassen and Biebach 1994)?

The elevated BMR associated with migratory disposition in many birds suggests that a general elevation of metabolism supports an increased capacity for endurance flight. Most likely, BMR tracks changes in the total aerobic capacity of migrants, which presumably increases in association with the high and sustained energy demands of migration (Vezina et al. 2007; Weber and Piersma 1996). In this respect, increases in BMR with migratory disposition do not directly promote enhanced endurance flight, but instead are indicative of adaptation of total aerobic capacity (Piersma 2002; Piersma et al. 1995, 1996a). That is, migration-induced increases in BMR result from increased maintenance costs of tissues involved in support (e.g., heart, gut, kidney) of the high energetic demands of peripheral effector organs (i.e., skeletal muscles) during migration (Piersma et al. 1996a, 1999). Moreover, as mass declines (along with flight and maintenance costs) during long-distance migratory flights, BMR may concurrently decline (Battley et al. 2000; Klaassen and Biebach 1994).

3.3.4 M_{sum} and Seasonal Acclimatization

A general pattern emerging from seasonal studies of both M_{sum} and cold tolerance in birds is that those species showing marked winter enhancement of cold tolerance also show significant increases in M_{sum} (Arens and Cooper 2005a; Liknes et al. 2002; Cooper 2002; Liknes and Swanson 1996; O'Connor 1995a; Cooper and Swanson 1994; Swanson 1990a; Marsh and Dawson 1989; Dawson and Smith 1986). Cold acclimation also increased M_{sum} and cold tolerance in Red Knots (Vezina et al. 2006). This suggests that M_{sum} and cold tolerance in birds are positively associated. Consistent with such a relationship, birds exhibiting only minor seasonal adjustment of cold tolerance also demonstrate little or no seasonal variation in M_{sum} (Swanson and Weinacht 1997; Dawson et al. 1983b). However, geographic variation in cold tolerance in small birds is not necessarily congruent with geographic variation in M_{sum}, suggesting that M_{sum} is a rather imprecise indicator of cold tolerance (Swanson 1993; Dawson et al. 1983b). Moreover, significant improvement of cold tolerance in winter in small birds can occur without or with only minor winter elevation of M_{sum} (Saarela et al. 1989b, 1995; Dawson and Smith 1986; Dawson et al. 1983b).

Marsh and Dawson (1989) and Dawson and Marsh (1989) reviewed the literature to that date on seasonal variation in M_{sum} and cold tolerance in small birds. They concluded that the major physiological adjustment associated with winter acclimatization in small birds was a seasonal increase in shivering endurance, rather than changes in thermogenic capacity. However, changes in endurance are associated with changes in aerobic capacity in vertebrates generally (Bennett 1991), so the relatively minor seasonal changes in M_{sum} accompanying seasonal changes in cold tolerance (i.e., shivering endurance under cold exposure) in birds seem somewhat unusual in this regard. Marsh and Dawson (1989) and Dawson and Marsh (1989) also noted that seasonal variation in M_{sum} in most birds amounted to less than about 15% and that such seasonal changes were lower in birds than in mammals, which sometimes exceeded 50%.

Much of the data used to formulate the conclusion of relatively minor seasonal adjustments in M_{sum} in small birds came from studies of cardueline finches, including the well-studied American Goldfinch *Carduelis tristis* (Dawson and Smith 1986; Dawson and Carey 1976) and the House Finch *Carpodacus mexicanus* (Dawson et al. 1983b). Dawson and Carey (1976) found no significant seasonal difference in mass-specific peak metabolic rates in goldfinches exposed to less than −60°C in air, despite winter birds tolerating these temperatures for up to 8 h while summer birds became hypothermic in less than 1 h. However, whole-organism peak metabolic rates in winter in these goldfinches were 32% higher than in summer. Dawson and Smith (1986) argued persuasively that whole-organism metabolic rates are actually the more pertinent measure for seasonal comparisons, largely because of seasonal differences in fattening. These authors detected a 16% elevation of M_{sum} in goldfinches in winter relative to spring, but M_{sum} in spring birds was measured in April, which has colder temperatures than summer months, so 16% might underestimate the full degree of seasonal variation in M_{sum} in these birds. Indeed, in American Goldfinches from South Dakota, whole-organism M_{sum} in winter exceeds that in summer by 31% (Liknes et al. 2002), a value similar to the whole-organism seasonal difference in peak metabolic rate documented by Dawson and Carey (1976). Thus, American Goldfinches appear to exhibit a greater degree of seasonal variation in M_{sum} than previously credited them. Dawson et al. (1983b) found no seasonal differences in M_{sum} from House Finches in California or Colorado, but O'Connor (1995a) found a 30% elevation of whole-organism M_{sum} in winter relative to summer in House Finches from Michigan. In addition, recent studies on a wider diversity of avian taxa have revealed that relatively large seasonal changes in M_{sum} are common among small birds, with winter increments of M_{sum} relative to summer commonly exceeding 25%, and exceeding 50% in some species (Liknes and Swanson 1996; Table 3.3). Thus, winter elevations of M_{sum} in birds and mammals appear less disparate than previously recognized. Finally, M_{sum} may also vary among winters or even within a winter in small birds and such variation is inversely correlated with temperature (Swanson and Olmstead 1999). This suggests that temperature exerts a proximate influence on M_{sum} in these birds.

The general positive relationship between M_{sum} and cold tolerance in birds suggests that the physiological changes underlying increases in M_{sum} may be important to winter acclimatization. M_{sum} cannot be sustained indefinitely, as birds become hypothermic after some time at a cold exposure eliciting peak thermogenesis (e.g., O'Connor 1995a). However, some fraction of M_{sum} can be maintained indefinitely, allowing survival of prolonged cold periods where energy demands for thermogenesis are high (Marsh and Dawson 1989). The fraction of M_{sum} that can be sustained for prolonged periods may be related to the relative provision of ATP to shivering muscles from fats or carbohydrates or from aerobic or anaerobic metabolism. In mammals, carbohydrates play an increasingly important role in ATP provision as exercise intensity increases, and lactate accumulates at high

Table 3.3 Variation in M_{sum} associated with seasonal acclimatization in birds

	Mass (g)		M_{sum} (mL O_2·min^{-1})			
Species	Summer	Winter	Summer	Winter	W/S	References
American Goldfinch (Michigan)	12.8	14.4	4.20	4.83	1.15	Dawson and Smith 1986
(S. Dakota)	12.4	13.8	4.27	5.61	1.31	Liknes et al. 2002
House Finch (California)	21.0	21.6	5.34	5.83	1.09[a]	Dawson et al. 1983b
(Colorado)	21.8	21.7	6.54	6.73	1.03[a]	–
(Michigan)[b]	21.1	21.6	5.46	7.12	1.30	O'Connor 1995a
(S. Dakota)	21.3	21.0	6.03	6.54	1.08[a]	Swanson and Liknes 2006
Eurasian Siskin	12.8	13.2	4.18	4.53	1.08	Saarela et al. 1995
Eurasian Greenfinch	27.1	26.6	6.19	5.19	0.84[a]	–
Evening Grosbeak	58.0	62.1	12.33	14.50	1.18	Hart 1962
European Starling	79.6	89.8	17.33	18.83	1.09	–
House Sparrow (Ontario)	27.1	29.4	7.0	10.0	1.43	–
(Wisconsin)	26.4	27.6	8.4	10.9	1.29	Arens and Cooper 2005b
(S. Dakota)	26.8	27.1	8.4	9.3	1.11	Swanson et al. 2009
Black-capped Chickadee	12.4	13.4	4.78	6.49	1.36	Cooper and Swanson 1994
Black-capped Chickadee	13.2	13.0	4.39	6.00	1.37	Swanson and Liknes 2006
Dark-eyed Junco	16.9	18.2	5.78	7.39	1.28	Swanson 1990a
White-breasted Nuthatch[d]	19.8	21.3	5.18	8.04	1.55	Liknes and Swanson 1996
White-breasted Nuthatch	19.6	21.8	4.87	6.21	1.28	Swanson and Liknes 2006
Downy Woodpecker	25.0	26.6	6.62	10.04	1.52	Liknes and Swanson 1996
Downy Woodpecker	25.8	26.0	7.73	7.19	0.93[a]	Swanson and Liknes 2006
Northern Bobwhite	220	230	20.19	22.02	1.09[a]	Swanson and Weinacht 1997
Mountain Chickadee	11.4	11.3	4.26	5.37	1.26	Cooper 2002
Juniper Titmouse	16.3	17.1	5.23	5.73	1.10	–
Northern Cardinal	41.4	48.3	8.10	10.91	1.35	Swanson and Liknes 2006

The W/S column is the ratio of M_{sum} in winter divided by M_{sum} in summer

[a]Seasonal values did not differ significantly

[b]Michigan House Finches were tested at −10°C in helox, so values may not represent actual summit metabolism. However, all birds became hypothermic at the end of these tests suggesting that metabolic rates closely approached summit metabolism

[c]Values from Hart (1962) are reported as presented in Marsh and Dawson (1989)

[d]Summer values for nuthatches are mid-summer values, which were significantly lower than early/late summer values (20.5 g, 6.20 mL O_2·min^{-1}) in this study

exercise intensities (e.g., 85% of maximum aerobic capacity), indicating an anaerobic contribution to ATP provision at these levels of exercise (Roberts et al. 1996), but this does not appear to be the case for birds, where fat serves as the major fuel for both shivering and locomotion at all levels of intensity (Vaillancourt et al. 2005; McWilliams et al. 2004).

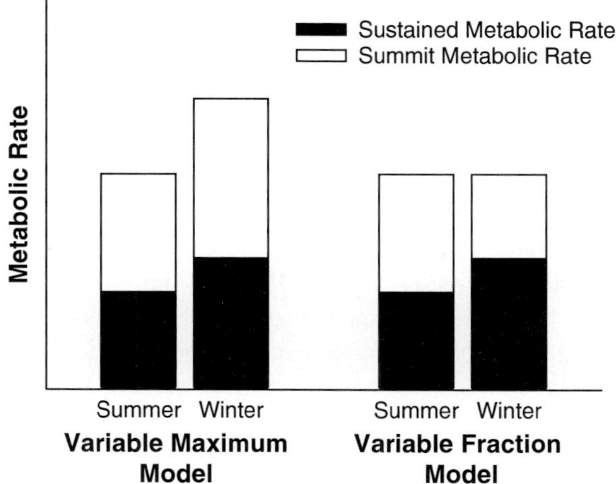

Fig. 3.3 Variable fraction and variable maximum models for how shivering endurance and M_{sum} are related on a seasonal basis. The entire extent of the bars respresents M_{sum}, while the filled portions of the bars indicate the metabolic rates that can be sustained indefinitely (some percentage of M_{sum}). The variable fraction model suggests that winter birds can sustain a higher fraction of M_{sum} than summer birds, thereby increasing shivering endurance at cold temperatures and improving cold tolerance. The variable maximum model suggests that M_{sum} varies seasonally and that sustaining a seasonally constant fraction of M_{sum} elevates sustainable levels of thermogenesis and thereby improves cold tolerance. The two models are not mutually exclusive. From Liknes et al. (2002)

Birds could presumably increase cold tolerance (i.e., shivering endurance) in winter by increasing the fraction of M_{sum} that could be sustained relative to that in summer (variable fraction model, Fig. 3.3). Alternatively, birds could sustain a seasonally constant fraction of M_{sum} but increase M_{sum} in winter (variable maximum model, Fig. 3.3). These two models explaining seasonal differences in shivering endurance are not mutually exclusive and the end result of both models is an increase in sustained heat production in winter relative to summer (Liknes et al. 2002). For mammals, interspecific differences in endurance are mediated by differences in aerobic capacity, rather than by differences in the fraction of maximum aerobic metabolic rate that can be sustained (Roberts et al. 1996). Marsh and Dawson (1989) suggested that because seasonal changes in M_{sum} in birds tested to that date were relatively minor (mostly <15%), winter increments of cold tolerance and shivering endurance were mediated through the ability of winter birds to sustain higher fractions of M_{sum} than summer birds, consistent with the variable fraction model. Recent data indicating larger seasonal variations in M_{sum}, however, are more consistent with the variable maximum model (e.g., Liknes et al. 2002). According to this model, species sustaining a seasonally constant fraction of M_{sum} would increase heat production in winter relative to summer for any given fraction of M_{sum} sustained. If we assume that shivering endurance is seasonally stable for any given fraction of M_{sum} sustained, a winter increase in M_{sum} would affect an increase in cold tolerance because sustainable levels of heat production would be higher in winter than in summer. Such an increase in sustained heat production would seem to be the functionally significant feature of winter increments of M_{sum} to cold tolerance and acclimatization (Liknes et al. 2002; Liknes and Swanson 1996; O'Connor 1995a). Furthermore, if a higher fraction of M_{sum} could be maintained in winter than in summer, as Marsh and Dawson (1989) suggest, then this could further improve cold tolerance in winter birds.

Because large winter increases in cold tolerance in birds may occur with only relatively minor winter increases in M_{sum} (Dawson and Smith 1986; Dawson and Carey 1976; Hart 1962) and intraspecific geographic variation in M_{sum} does not necessarily parallel geographic variation in cold tolerance, the precise nature of the relationship between cold tolerance (i.e., increased shivering endurance) and M_{sum} in birds is uncertain. Swanson (2001) demonstrated that mass-independent M_{sum} and shivering endurance (measured as time to hypothermia under a standardized cold stress in helox) were positively correlated for Dark-eyed Juncos (*Junco hyemalis*), American Tree Sparrows (*Spizella arborea*), and Black-capped Chickadees (*Poecile atricapillus*) overwintering in South Dakota, although the variation in M_{sum} explains only a portion of the variance in endurance ($R^2 = 0.11$–0.54). Moreover, mass-adjusted M_{sum} was positively associated with cold tolerance on an interspecific basis for 25 bird species (Swanson and Liknes 2006). Thus, the general positive correlation between M_{sum} and cold tolerance (i.e., shivering endurance) in birds is consistent with the general rule among vertebrates of a positive association between expanded aerobic capacities and enhanced endurance at submaximal levels of activity (Bennett 1991; Marsh and Dawson 1989).

3.3.5 M_{sum} and Migratory Disposition

Because physiological adjustments for migration and winter acclimatization both promote endurance muscular activities (long-distance flight and prolonged shivering, respectively) and organismal M_{sum} generally increases in winter, it might be expected that migratory disposition could influence M_{sum}. Seasonal variation in M_{sum} associated with migration has been documented in four passerine migrants (Table 3.4, Swanson and Dean 1999; Swanson 1995). For Warbling Vireos *Vireo gilvus* and Yellow Warblers *Dendroica petechia*, per-bird M_{sum} was elevated (18 and 23%, respectively) during spring migration relative to summer and fall migration periods, which do not differ significantly. M_{sum} in Ruby-crowned Kinglets *Regulus calendula* and Yellow-rumped Warblers *Dendroica coronata* was measured only during spring and fall migration periods and was significantly greater in spring than in fall for warblers (20%) and for male kinglets (11%), but not for female kinglets. The general pattern emerging from these studies is that M_{sum} is elevated during spring migration, relative to other seasons, at least in the territory-establishing sex. Interestingly, the degree of seasonal variation in M_{sum} associated with migration shows broad overlap with the percent increases in M_{sum} associated with winter

Table 3.4 Variation in M_{sum} associated with migratory disposition in passerine birds

Species	Mass (g)			M_{sum} (mL $O_2 \cdot min^{-1}$)			S/F[a]	References
	Spring	Sum	Fall	Spring	Sum	Fall		
Warbling Vireo	14.4	13.6	14.1	5.00	4.19	4.26	1.18	Swanson 1995
Ruby-crowned Kinglet (male)	6.3	–	6.0	2.63	–	2.37	1.11	Swanson and Dean 1999
Ruby-crowned Kinglet (female)	5.6	–	5.7	2.17	–	2.14	1.01 (NS)	–
Yellow Warbler	10.0	9.1	9.5	3.46	2.89	2.73	1.23	–
Yellow-rumped Warbler	12.7	–	11.8	4.49	–	3.75	1.20	–

[a]Ratio calculated as spring M_{sum} divided by fall M_{sum} or pooled fall and summer M_{sum}

acclimatization (0–55%, but most values <35%, Table 3.3). In addition to these studies on passerines, Vezina et al. (2007) found that M_{sum} and cold tolerance in captive Red Knots changed in tandem with body mass and pectoralis muscle size changes over the period encompassing spring migration, and that this relationship was independent of thermal acclimation. More studies measuring seasonal trends in M_{sum} associated with migratory disposition on a wider variety of taxa are needed. Specific research questions in need of testing include: (1) Is the general pattern so far documented for passerines and Red Knot robust? (2) Do fall migratory adjustments in M_{sum} generally differ from those during spring or does variation in M_{sum} relate to differing migration strategies? (3) Are the percent changes in M_{sum} associated with migration and winter acclimatization similar? (4) Does M_{sum} change with the duration of migration (i.e., approaching the migratory destination), as rates of migration sometimes do (Ellegren 1993)?

The potential significance for the general pattern of a spring increment in M_{sum} in migratory birds is less obvious than the increase in M_{sum} associated with winter acclimatization because most migrants are not exposed to the seasonal extremes of temperature encountered by residents of cold climates. Swanson (1995) and Swanson and Dean (1999) presented two hypotheses, the cold acclimatization hypothesis and the flight adaptation hypothesis, to explain the pattern of seasonal variation in M_{sum} associated with migration documented for passerine migrants. The cold acclimatization hypothesis posits that selection for improved cold hardiness during spring migration relative to other seasons is responsible for the spring increment of M_{sum}. Many migrants are subject to colder temperatures during spring migration and upon arrival on breeding grounds than during the remainder of the breeding season or during fall migration (Swanson and Dean 1999; O'Reilly and Wingfield 1995; Wiersma and Piersma 1994). Thus, provided that increases in M_{sum} are related to increases in cold hardiness (see above), elevated M_{sum} and cold resistance during spring could be beneficial from a thermoregulatory standpoint.

Alternatively, the flight adaptation hypothesis suggests that the spring increment of M_{sum} is a by-product of selection for improved endurance flight during the spring rush to the breeding grounds. Faster flight velocities, longer flight durations, or shorter stopover periods between successive flights could all produce a faster pace of migration in spring than in fall. The intensity, duration, and frequency of exercise training appear to be important in eliciting metabolic adjustments associated with improved muscular endurance in mammals and/or birds (Butler and Turner 1988; Harms and Hickson 1983; Hickson 1981). Thus, behaviors resulting in a faster pace of migration might be expected to produce improved flight endurance in migrating birds.

For the flight adaptation hypothesis to be a valid explanation for the patterns of seasonal variation in M_{sum} associated with migration documented for passerines by Swanson (1995) and Swanson and Dean (1999), spring rates of migration must exceed rates of migration in fall. This appears to be the case for several Old World passerine migrants and spring migration rates can exceed fall migration rates by two- to threefold in these species (Fransson 1995; Pearson and Lack 1992). Data are not available to test whether seasonal differences in migration rates exist for New World migrants, but some indirect evidence suggesting a faster pace of migration in spring than in fall has been reported (Swanson and Dean 1999; Morris et al. 1994; Winker et al. 1992).

One method of testing whether M_{sum} variation associated with migration conforms better to cold acclimatization or flight adaptation hypotheses is to examine seasonal variation in M_{sum} in migrants with differing migration strategies. Swanson and Dean (1999) used this method to compare seasonal variation M_{sum} in the mild temperate-zone migrants, Ruby-crowned Kinglet and Yellow-rumped Warbler, with that in the Neotropical migrants, Warbling Vireo, and Yellow Warbler. Mild temperate-zone migrants migrate earlier in the spring and later in the fall than Neotropical migrants (Hagan et al. 1991), so they are exposed to lower average and extreme temperatures, increased probabilities of encountering cold or adverse weather, and are subject to reduced variation between spring and fall migration periods than are Neotropical migrants. Swanson and Dean (1999) found that mild temperate and Neotropical migrants demonstrated similar mass-independent M_{sum} and similar levels of seasonal variation in M_{sum}, despite the differences in thermal environments encountered. These data are not consistent with the cold acclimatization hypothesis and suggest that factors other than temperature acclimatization, perhaps adaptation for endurance flight, are responsible for the migration-induced changes in M_{sum}. The data of Vezina et al. (2007) for captive Red Knots provide a more direct test of the flight adaptation hypothesis and the increases in M_{sum} and cold tolerance during the period of spring migratory fattening and pectoralis muscle hypertrophy, irrespective of thermal acclimation, also support the idea that elevated thermogenic capacity is a by-product of migratory disposition. Even if cold acclimatization is not the selective factor responsible for the spring increment of M_{sum} in migratory birds, elevated M_{sum} and accompanying increases in cold tolerance could still presumably benefit thermoregulation in these birds if they encounter cold temperatures or adverse climatic conditions during migration or upon arrival on breeding grounds. However, the spring increment of M_{sum} in the passerine migrants studied by Swanson (1995) and Swanson and Dean (1999) was not uniformly associated with improved cold resistance.

3.4   Mechanistic Correlates of Seasonal Variation in Metabolic Rates

Adjustments contributing to seasonal changes in whole-organism basal, summit, or MMRs could occur at several levels within the bird (Fig. 3.4). BMR is largely a function of central organs (e.g., brain, heart, gut, and kidney), while maximal and summit metabolic rates are largely functions of skeletal muscle activity (Hoppeler and Weibel 1998; Bennett 1991; Taigen 1983). Consequently, adjustments promoting changes in MMR or M_{sum} might not promote similar changes in BMR. Thus, M_{sum} and BMR may not show corresponding seasonal or migration-induced variation in birds. Such a discrepancy between seasonal variation in BMR and M_{sum} has been demonstrated in some birds (O'Connor 1995a; Dawson and Smith 1986; Hart 1962). On the other hand, BMR and M_{sum} are phenotypically correlated in birds independent of mass and phylogeny (Rezende et al. 2002; Dutenhoffer and Swanson 1996; but see Wiersma et al. 2007b), so physiological and biochemical adjustments influencing M_{sum} in birds may generally influence BMR or vice versa.

Because seasonal or migration-induced changes in BMR in birds apparently reflect increased maintenance costs for support of elevated metabolic capacities (M_{sum} and MMR) (Piersma et al. 1995, 1996a; Swanson 1991a), a review of potential limits to aerobic metabolism is needed before examination of mechanistic responses underlying changes in BMR and metabolic capacities can proceed. Organismal metabolic capacities are potentially limited at several steps along the

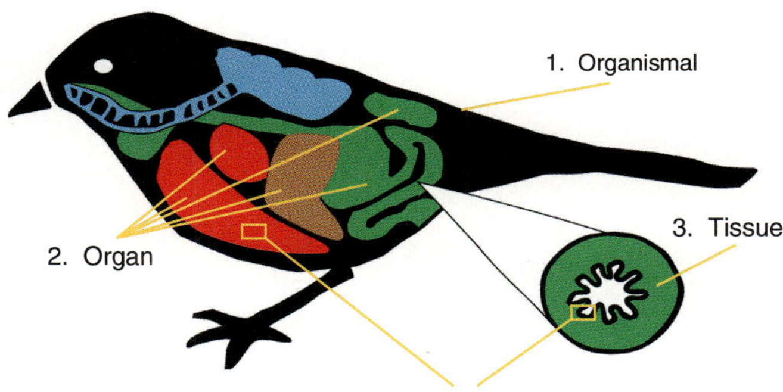

Fig. 3.4 Adjustments potentially mediating seasonal variation in basal, summit, or MMRs can occur from the whole-organism to the biochemical levels within the bird. At the whole-organism level, changes in fattening or thermal conductance, along with changes in organismal metabolic rates, could markedly influence flight or shivering endurance. Such organism-level adjustments in basal or MMRs could themselves be mediated by changes in masses of metabolically active nutritional or exercise organs. Variation at the cell or tissue level, such as cellular hypertrophy or changes in capillary density in the muscle or lumenal surface area of the intestine, could complement organ-level adjustments to enhance metabolic or nutritional support capacities. Finally, contributions from biochemical changes, such as increased catabolic enzyme activities, increased capacities of metabolic substrate transport, or increased intestinal transporter uptake rates, could influence mass-specific metabolic intensities of organs and thereby impact organismal metabolic rates

pathways for provision and use of oxygen and metabolic substrates (Hoppeler and Weibel 1998; Suarez 1998). Limits might be imposed by oxygen diffusing capacity in the lung, substrate mobilization from storage depots, transport capacity of oxygen and substrates to tissues, capillarity of muscles, transport of oxygen and substrates into muscle cells, intracellular transport to mitochondria, availability of intracellular

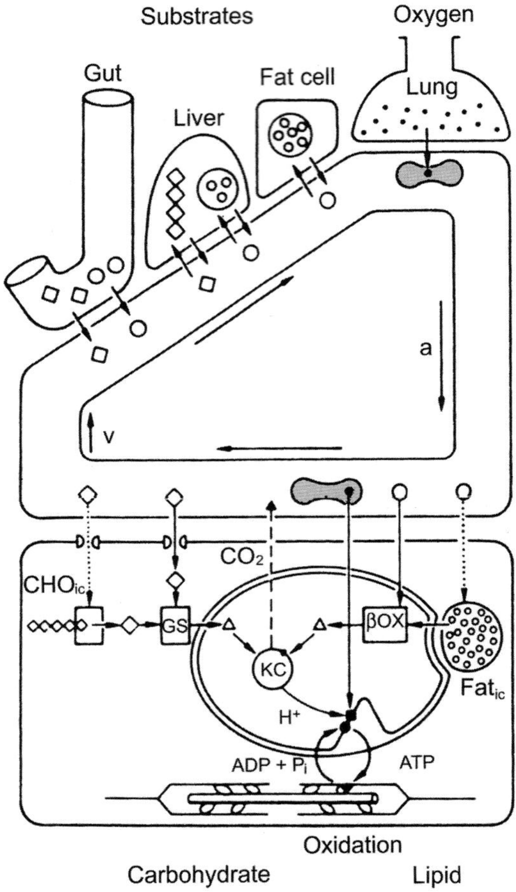

Fig. 3.5 Diagram illustrating the interacting structures and pathways involved in the processing, transport, and use of oxygen and substrates during oxidative metabolism. Substrates enter the bloodstream from the gut or from storage depots, while oxygen enters at the lung. Both oxygen and substrates are delivered to the muscle cell and are transported into the mitochondria for oxidative metabolism. Limitations to aerobic capacity might be imposed by oxygen diffusing capacity in the lung, substrate mobilization from storage depots, transport capacities of oxygen or substrates to muscle cells, capillarity of muscles and transport of oxygen and substrates into muscle cells, intracellular transport to mitochondria, availability of intracellular substrate stores to supplement extracellular supply, or cellular oxidative capacity. *CHO* carbohydrates; *GS* glycolysis; *KC* Krebs cycle; *βOX* β-oxidation; *Fat_ic* intracellular lipid deposits; *a* arterial; and *v* venous. *Open circles* indicate lipids, *open squares* indicate carbohydrates, and *closed dots* indicate oxygen. Reprinted from Hoppeler and Weibel (1998), with permission

substrate stores to supplement extracellular supply, or cellular energy use (Fig. 3.5). Interspecific studies of mammals suggest that the following factors are important to MMRs: lung volume, stroke volume of the heart and blood oxygen carrying capacity, capillarity of muscles, transport of oxygen into muscle cells, intracellular substrate supplies in the muscle, and muscle mitochondrial volume (Hoppeler and Weibel 1998). Thus, limitations to aerobic capacity appear at a number of levels and limits may vary under differing conditions for activity. This suggests that the capacities for any one level of the integrated system do not greatly exceed capacities at other levels, a concept known as symmorphosis (Taylor and Weibel 1981). Symmorphosis, or at least correlated changes in some steps of pathways affecting organismal performance, appears to be generally applicable to limits on aerobic performance in vertebrates, including birds (Seymour et al. 2008; Suarez 1998; Bennett 1991). Thus, elevated organismal aerobic capacity is often accompanied by changes in several components of energy processing, transport, and use.

Nevertheless, metabolism of the skeletal muscle cells determines the aerobic energy demand during maximal metabolism (Hoppeler and Weibel 1998; Bennett 1991). With regard to energy demand, seasonal or migration-induced variation in summit metabolic rates could conceivably result from either (or both) changes in metabolic intensity (i.e., mass-specific alteration of metabolism) or masses of muscles active in shivering. Factors involved in regulating muscle metabolic intensity include mitochondrial volume, mitochondrial cristae surface area, and concentrations of catabolic enzymes (Suarez 1998). The flight muscles (pectoralis and supracoracoideus) are probably primary determinants of MMRs during both migratory flights and prolonged shivering, but the leg muscles may also be important for shivering in some species (Vittoria and Marsh 1996; Duchamp and Barre 1993; Marsh and Dawson 1989; Carey et al. 1989; Aulie and Tøien 1988). Thus, if variation in metabolic intensity is associated with changes in aerobic capacity, it should occur at these sites. Such variation would alter the metabolic capacity of skeletal muscles without corresponding changes in muscle mass. Regulatory enzymes especially important to control aerobic capacity include citrate synthase (CS), a regulatory enzyme of the Krebs cycle, and cytochrome c oxidase (CCO), which is the terminal step in the electron transport chain. Activities of both of these enzymes have been used as indices of mass-specific aerobic capacity of tissues (e.g., Vezina and Williams 2005; Liknes 2005; Weber and Piersma 1996; O'Connor 1995b; Marsh and Dawson 1982). Enzyme activities typically used as indicators of carbohydrate and lipid provision capacities to the Krebs cycle in birds are phosphofructokinase-1 (PFK-1), a regulatory step for glycolysis, and β-hydroxyacyl-CoA dehydrogenase (HOAD), a regulatory step in the β-oxidation pathway for catabolism of free fatty acids (FFA) (e.g., Liknes 2005; Guglielmo et al. 2002; O'Connor 1995b; Marsh and Dawson 1982; Marsh 1981).

Whole-organism variation in aerobic metabolic capacity could also occur without changes in metabolic intensity by variation in the masses of skeletal muscles (Daan et al. 1990). During migratory periods or season-long cold exposure, increases in energy demand resulting from increased mass and activity of skeletal muscles would also entail increased processing and provision of energy to support elevated metabolic rates. The heart, gut, digestive organs, and kidney might undergo seasonal or migration-induced adjustment to meet these heightened energy demands and such changes in these central organs would presumably influence BMR.

3.4.1 Seasonal or Migration-Induced Changes in Transport Capacities for Oxygen

The principle of symmorphosis dictates that elevated energy demand by the skeletal muscles during migration or winter should be accompanied by increases in other systems involved in energy provisioning. Critical steps in the provision of energy to active muscles involve the transport and delivery of oxygen and substrates to muscles. Seasonal or migration-induced variation in metabolic rates would therefore be expected to be associated with alterations in oxygen and substrate transport mechanisms. Oxygen carrying capacity of the blood is determined by the hemoglobin concentration, which in turn is related to the hematocrit, or packed cell volume, of the blood. Hematocrit was elevated by 11% and oxygen carrying capacity by 9% in Dark-eyed Juncos in winter relative to summer (Swanson 1990b). In White-crowned Sparrows, hematocrit was elevated in winter and during spring migration relative to other periods of the year (deGraw et al. 1979). Moreover, migratory Bar-tailed Godwits (*Limosa lapponica*) had higher hematocrit and hemoglobin content than allometrically predicted, and both increased in birds that improved their energetic condition during stopover, suggesting that high levels are associated with readiness for departure on migratory flights (Landys-Ciannelli et al. 2002; Piersma et al. 1996b). Hematocrit was also positively correlated with MMR during treadmill running in female, but not male, Red Junglefowl *Gallus gallus* (Hammond et al. 2000). In contrast, Breuer et al. (1995) found that four species of Australian passerines all showed seasonally constant hematocrit, but winter increases in erythrocyte number, suggesting smaller erythrocytes with greater surface area in winter relative to summer. These authors speculated that the smaller erythrocytes in winter could increase efficiency of gas exchange and oxygen transport in support of winter thermogenic demands. Hematocrit also showed little seasonal variation in captive Ruff *Philomachus pugnax* and Red Knot during the annual cycle, despite maintenance of typical annual cycles of body mass and molt in these birds (Piersma et al. 2000). In a recent review of hematocrit variation in birds, Fair et al. (2007) concluded that hematocrit generally, but not universally, increased in winter relative to summer for birds in temperate climates.

Transport of oxygen from blood to working muscles is also important to aerobic muscle function and depends, among other factors, on the vascularization of the muscle (high vascularity decreases diffusion distance) and the gradient for diffusion. In addition, oxygen transfer from blood to muscle can be facilitated by muscle myoglobin because its affinity for oxygen is higher than that of hemoglobin. In an interspecific study of pectoralis muscle capillary density in 15 species of European passerines, Lundgren and Kiessling (1988) found that capillary density was higher in long-distance migrants when compared with short-distance migrants or residents, suggesting that increased capillary density is important to endurance flight. Swim training increased both the maximum oxygen consumption during swimming (27%) and the capillary to muscle fiber ratio (20%) in leg muscles of Tufted Ducks *Aythya fuligula* (Butler and Turner 1988). Cold acclimation in the pigeon increased capillary density in pectoralis muscle, but no such change occurred between flying and sedentary birds (Mathieu-Costello et al. 1994, 1998). Treadmill training in Bar-headed Geese *Anser indicus* increased leg muscle myoglobin content by 20–31% in different thigh muscles (Saunders and Fedde 1991). Captive Tufted Ducks in an outdoor aviary that underwent a natural training regimen that included diving and flying had higher myoglobin content in pectoralis and leg muscles (38–55%) than untrained indoor captives (Butler and Turner 1988). In addition, European Starlings had higher myoglobin content in both heart and pectoralis muscle shortly after arrival from migration when compared to wintering birds in Spain (Palacios et al. 1984). In general, both the capacity of the blood to transport oxygen and the ability to unload oxygen to the tissues seem to increase with activity training, migration, or cold exposure in birds. In contrast to these results, however, American Goldfinches showed no difference in capillary to muscle fiber ratios in the pectoralis between winter and late spring (Carey et al. 1978).

Oxygen extraction efficiency (EO_2) may vary under cold exposure in birds and seasonal changes in EO_2 could contribute to variation in oxygen delivery to working muscles among seasons (Arens and Cooper 2005a; 2005b). Only two species have been studied to date with regard to seasonal changes in EO_2. Seasonally stable EO_2 under severe (helox) and moderate cold exposure was documented for House Sparrows during the active phase of the daily cycle, but moderate cold exposure resulted in higher EO_2 in winter than in summer during the rest phase, when thermogenic demands are typically greatest (Arens and Cooper 2005a, b). Black-capped Chickadees showed winter increases in EO_2 relative to summer under severe cold stress (Cooper and Same 2000).

3.4.2 *Seasonal or Migration-Induced Changes in Transport Capacities for Substrates*

Prolonged shivering and endurance flights are both fueled largely by lipid (Dawson et al. 1983a) and birds rely on circulatory lipids to fuel sustained exercise to a greater degree than mammals (McWilliams et al. 2004; Weber et al. 1996), so transport capacities for lipids might be expected to increase with winter acclimatization or migratory disposition. Because lipids are relatively insoluble in plasma, they are generally carried in the blood bound to carrier proteins. FFA are generally bound to plasma albumin and triglycerides (TG) are bound to apoproteins, mainly as very low-density lipoproteins (VLDL) for transport to muscle (Fig. 3.6; McWilliams et al. 2004; Stevens 1996; Ramenofsky 1990). Plasma FFA levels are generally

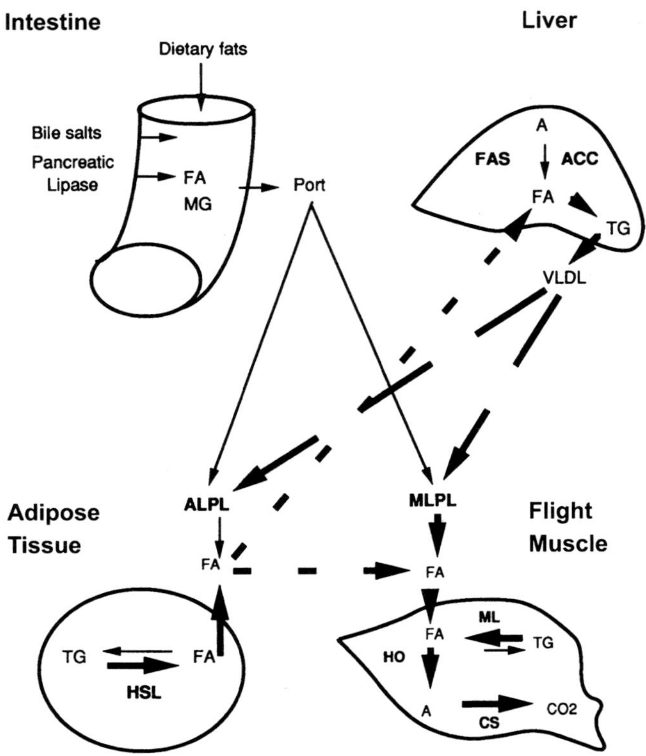

Fig. 3.6 Pathways for lipid mobilization and use as a metabolic substrate for migration. The *heavy lines* indicate major metabolic pathways operating during migration. Lipids are mobilized from stored fat and transported to liver and flight muscle as free fatty acids (FA) bound to plasma albumin. Triglycerides (TG) from the liver may also be released into the blood and bound to apo-proteins to form very low-density lipoproteins (VLDL). VLDLs can be hydrolyzed by muscle lipoprotein lipase (MLPL) for transfer into the muscle cell as FA. Enzymes are shown in *bold*. *MG* monoacylglycerol; *Port* portomicrons; *HSL* hormone-sensitive lipase; *ALPL* adipose lipoprotein lipase; *FAS* fatty acid synthethase; *ACC* acetyl-CoA carboxylase; *ML* muscle lipase; *HO* β-hydroxyacyl-CoA dehydrogenase; and *CS* citrate synthase. Reprinted from Ramenofsky (1990), with permission

elevated during cold exposure (Liknes 2005; O'Connor 1995b; Swanson 1991b; Marsh et al. 1984; Marsh and Dawson 1982) and sustained flights or hop-flutter wheel exercise (Pierce et al. 2005; Landys et al. 2005; Jenni-Eiermann et al. 2002; Gannes et al. 2001; Jenni-Eiermann and Jenni 1992, 2001) in birds, although plasma levels under such conditions often do not vary seasonally. Moreover, plasma levels of albumin did not change between winter and late spring in House Finches, despite improved cold tolerance and elevated M_{sum} in winter birds (O'Connor 1995b). Plasma TG and VLDL were elevated during active migratory flights when compared to resting or foraging birds in European Robins *Erithacus rubecula*, Pied Flycatchers *Ficedula hypoleuca*, and Garden Warblers *Sylvia borin* (Jenni-Eiermann and Jenni 1992). Moreover, Bar-tailed Godwits captured immediately following migratory flights had elevated TG levels compared to inactive fasting birds, also suggesting that elevated TG levels may support the high metabolic demands of flight (Landys et al. 2005). This raises the possibility that the transport of TG via VLDL might be a general mechanism for seasonal variation in lipid transport associated with migration or winter acclimatization in birds. However, TG levels did not show similar elevation with flight for pigeons (Schwilch et al. 1996; Bordel and Haase 1993) or Red Knots (Jenni-Eiermann et al. 2002). Neither did TG levels increase with hop-flutter wheel exercise in Red-eyed Vireos *Vireo olivaceous* (Pierce et al. 2005) or with severe cold exposure (in either summer or winter) in Black-capped Chickadees, House Sparrows, and White-breasted Nuthatches *Sitta carolinensis* (Liknes 2005). Thus, seasonal modulation of lipid supply to working muscles via TG and VLDL does not appear to be a common mechanism supporting high levels of muscular activity among birds.

Muscle lipoprotein lipase (MLPL), present in the capillary endothelial wall, catalyzes the hydrolysis of plasma TG to FFA and glycerol, which can then be taken up by the muscle cells (Ramenofsky 1990). Peak MLPL activity in captive Dark-eyed Juncos during the spring migratory period occurred when birds were exhibiting migratory restlessness, suggesting an increased supply of lipid to muscles during periods of migratory flight (Ramenofsky 1990). Seasonal changes in MLPL activity might allow modulation of uptake of lipids by muscles. Pectoralis lipoprotein lipase activity, however, did not vary between premigratory and wintering Rosy Pastors (*Sturnus roseus*) from India (George and Vallyathan 1964). Savard et al. (1991) also failed to document any difference in peak MLPL activity between migratory and nonmigratory juncos. Furthermore, mean MLPL activity was higher in outdoor captive Dark-eyed Juncos during early molt than during migration (Ramenofsky et al. 1999). On the other hand, diel rhythms of MLPL activity in juncos were apparent during migration, but not during winter or molt, with nocturnal values associated with migratory restlessness greater than diurnal values (Ramenofsky et al. 1999). MLPL activity declined overnight in juncos exhibiting nocturnal migratory restlessness, concomitant with an increase in adipose tissue lipolysis, suggesting a relatively steady supply of FFA to muscles during nocturnal activity (Savard et al. 1991). Thus, although lipids are the primary fuel source for prolonged shivering and migratory flights, conclusive demonstration of enhanced capacities of circulatory lipid transport to working muscles during migration or winter in birds is lacking.

Another potential regulatory point for lipid metabolism is the delivery of fatty acids from the plasma across the sarcolemma into the muscle cell. Fatty acid transporters account for much of this delivery (McWilliams et al. 2004), so could function as sites of migration-induced or seasonal variation. Sarcolemmal fatty acid transporters in birds and mammals include plasma membrane-bound fatty acid binding protein (FABP$_{pm}$) and fatty acyl translocase (FAT/CD36) (Sweazea and Braun 2006; Bonen et al. 2004; Abumrad et al. 1999; Luiken et al. 1999). McFarlan (2007) recently studied migratory variation in sarcolemmal fatty acid transporters in White-throated Sparrows (*Zonotrichia albicollis*) and found that gene expression in the pectoralis muscle for both FABP$_{pm}$ and FAT/CD36 increased in migratory birds from Ontario relative to wintering birds from southern Mississippi, with fall migrants showing higher expression than spring migrants. Pectoralis $FABP_{pm}$ protein levels also increased in migratory sparrows relative to winter, but spring and fall protein levels did not differ. These data suggest that sarcolemmal fatty acid transporters may be an important regulatory site for phenotypic flexibility in overall fatty acid flux rates during prolonged muscular exercise in birds.

A limited amount of information is available regarding migration-induced variation in intracellular transport of lipids to the mitochondrial matrix, where β-oxidation of fatty acids occurs. In exercising mammals, maximum delivery of extracellular carbohydrate and lipid to the muscle cells occurs at moderate workloads, about 40–50% of MMR. Higher work rates require reliance on intracellular substrate stores, which are higher in athletic species than in sedentary species of mammals (Hoppeler and Weibel 1998). Heart-type fatty acid binding protein (H-FABP) facilitates the intracellular transport of fatty acids within the skeletal muscle cell (Pelsers et al. 1999). H-FABP is a prominent cytosolic protein in pectoralis muscle and heart of migrating Western Sandpipers *Calidris mauri* where it is found at levels several-fold higher than in nonvolant mammals (Guglielmo et al. 1998). Guglielmo et al. (1998) further suggest that modulation of muscle H-FABP expression could serve as an important mechanism for enhancing extracellular lipid use during migratory flights. In Barnacle Geese *Branta leucopsis* pectoralis H-FABP increases throughout development and is further increased, in wild birds but not in captives, just prior to migration suggesting a relationship between H-FABP and aerobic capacity (Pelsers et al. 1999). In addition, increases in pectoralis H-FABP and/or H-FABP expression have been documented during migration for Western Sandpipers (Guglielmo et al. 2002) and White-throated Sparrows (McFarlan 2007) and during winter for Black-capped Chickadees and White-breasted Nuthatches, although H-FABP remained seasonally stable in House Sparrows (Liknes 2005). H-FABP was also seasonally stable in the supracoracoideus muscle for Black-capped Chickadees, White-breasted Nuthatches and House Sparrows (Liknes 2005). These data suggest that intracellular lipid transport via H-FABP in the pectoralis is generally elevated by the increased metabolic demands of winter and, particularly, migration, so intracellular fatty acid transport may be another key regulatory site for phenotypic flexibility of metabolism in birds.

Few data are available to assess whether intracellular lipid stores in avian muscle are as important as they are in mammals for support of high aerobic workloads. The volume density of intracellular lipid droplets in pigeon pectoralis muscle approximately doubled after cold exposure compared to controls, but flight activity did not promote a similar change relative to sedentary controls (Mathieu-Costello et al. 1994, 1998). Intracellular lipid in the pectoralis muscle of Eared Grebes *Podiceps nigricollis* approximately doubled during hypertrophy following the flightless molt period and just prior to fall migration (Gaunt et al. 1990). Piersma et al. (1999) detected little change in the fat content of flight muscles during staging in red knots, despite marked changes in the size of the muscles, suggesting little change in intracellular fat content. Intramuscular lipid in pectoralis muscles of Black-capped Chickadees, White-breasted Nuthatches and House Sparrows from the cold winter climate of South Dakota was seasonally stable (Liknes 2005). Thus, available data are equivocal with regard to whether intracellular lipid in avian muscle increases with increasing energy demands. The absence of consistent increases in intracellular lipid with increasing energy demand is, perhaps, not surprising given the primary reliance of birds on exogenous fat stores to fuel sustained exercise (McWilliams et al. 2004).

The final step in lipid delivery to the mitochondria is transport of FFA across the mitochondrial membrane into the matrix where β-oxidation enzymes are housed. Transport of FFA into the mitochondrial matrix is catalyzed carnitine acyl CoA transferase (CAT), but may also involve FAT/CD36, which occurs on the mitochondrial membrane in mammals, may be physically associated with CAT, and functions in transport of long-chain fatty acids into the mitochondria in exercising muscle (Holloway et al. 2006; Campbell et al. 2004). In Semipalmated Sandpipers *Calidris pusilla* that were ready to depart on migration following a stopover period, carnitine oleoyl CoA transferase activity was significantly elevated relative to birds that were not in migratory disposition (Driedzic et al. 1993). Similarly, carnitine palmitoyl transferase increased in migratory relative to nonmigratory Western Sandpipers (Guglielmo et al. 2002). These data suggest that the capacity for FFA transport into the mitochondrion is enhanced with migratory disposition, but whether this result applies to other migrants or to birds wintering in cold climates is unknown. Based on the limited information available, it appears that intracellular transport of FFA to the mitochondrial matrix may vary with energy demand in birds, so this would seem to be a profitable avenue for further research relating to seasonal or migration-induced adjustments in metabolic rate.

3.4.3 Seasonal or Migration-Induced Changes in Mass-Specific Metabolic Intensity

Mass-specific adjustment of metabolic intensity could result from elevated mitochondrial density in tissues or from enhanced mass-specific activities of catabolic enzymes. Mitochondrial density in muscle as a function of cold acclimation, migration, or flight activity has only been investigated for a few species and these data suggest that mitochondrial density does increase positively with energy demand. Pectoralis muscle mitochondrial density increased concomitantly with muscular hypertrophy during the premigratory phase in three species of *Calidris* sandpipers (Evans et al. 1992). Mitochondrial density in pectoralis muscle also increased with hypertrophy in Eared Grebes during the period following molt and immediately prior to fall migration (Gaunt et al. 1990). Flight activity and cold acclimation both increased mitochondrial volume density in pigeon pectoralis muscle (Mathieu-Costello et al. 1994, 1998).

If mass-specific aerobic capacity of muscles increases during migration or winter acclimatization, such changes would need to be supported by an increased flux of substrates, principally lipid and carbohydrates, into the Krebs cycle. Thus, increased activities of enzymes in β-oxidation and glycolysis pathways might be expected with development of migratory disposition or with winter acclimatization. Winter acclimatized American Goldfinches showed elevated mass-specific activities of PFK-1 and HOAD in pectoralis muscle relative to their summer counterparts, but hexokinase (HK) activity (an indicator of catabolism of glucose from plasma) was seasonally stable (Yacoe and Dawson 1983; Marsh and Dawson 1982). In House Finches, mass-specific activities of pectoralis muscle HK, PFK-1, and HOAD were all seasonally stable, while leg muscle showed seasonally stable HK and PFK-1, but winter increases in HOAD (O'Connor 1995b; Carey et al. 1989). Black-capped Chickadees exhibited elevated mass-specific and total HOAD activities in winter compared to summer for pectoralis and leg muscles, but not for supracoracoideus, and HOAD activity was seasonally stable in all these muscles for White-breasted Nuthatches and House Sparrows (Liknes 2005). For the same three species, activity of PFK-1 was seasonally stable in pectoralis, supracoracoideus and leg muscles (Liknes 2005).

Guglielmo et al. (2002) documented a 12% increase in pectoralis HOAD activity during migration in Western Sandpipers. Elevation of mass-specific activities of pectoralis HOAD in migratory compared with nonmigratory individuals is commonly reported for passerine migrants (Lundgren 1988; Lundgren and Kiessling 1985, 1986; Marsh 1981), but such elevation does not occur universally. Migratory Sedge Warblers *Acrocephalus schoenobaenus*, Reed Buntings *Emberiza schoeniclus*, and Yellowhammers *Emberiza citrinella* showed seasonally constant activities of pectoralis HOAD (Lundgren 1988; Lundgren and Kiessling 1985). Glycolytic enzyme activities (PFK-1 or pyruvate kinase) in pectoralis are unchanged or decrease with migratory disposition in most species (Lundgren 1988; Lundgren and Kiessling 1985; Marsh 1981), but increases have been reported in migratory Semipalmated Sandpipers *Calidris pusilla* (Driedzic et al. 1993) and juvenile Reed Warblers *Acrocephalus scirpaceus* (Lundgren and Kiessling 1986). Generalizing from these data, an increased capacity for lipid oxidation in pectoralis muscle is common, but not universal, to both winter acclimatization and migration in birds, but alteration of carbohydrate oxidation capacity is less regularly associated with these conditions.

For alterations in substrate catabolism to be important to changes in mass-specific aerobic capacity, enzyme activities in Krebs cycle and oxidative phosphorylation pathways would also need to be adjusted to permit changes in flux through these pathways. Current data are equivocal with reference to whether seasonal or migration-induced variation in mass-specific activities of oxidative enzymes underlie changes in aerobic capacity, as some species show changes while others do not. Muscovy ducklings, *Anas barbariae*, exhibited increased activities of CS and CCO in pectoralis and hindlimb muscles after cold acclimation (Vittoria and Marsh 1996; Barré et al. 1987). Black-capped Chickadees and House Sparrows showed elevated CS activity in pectoralis and/or supracoracoideus muscles (but not in leg muscles) in winter relative to summer (Liknes 2005). Likewise, European Tree Sparrows *Passer montanus* showed elevated pectoralis CCO activity in winter compared to summer (Zheng et al. 2008). European

Starlings had higher mass-specifc and total pectoralis CS activity at the end of winter than during the egg-laying period, and total activity was also higher at the end of winter than during the chick-rearing period (Vezina and Williams 2005). Treadmill training elevated CS activity in leg muscles of the Tufted Duck relative to untrained individuals (Butler and Turner 1988). Combined pectoralis and leg muscle CS activity was positively correlated with MMR during treadmill running in male, but not female, Red Junglefowl (Hammond et al. 2000). Reed Warbler, European Robin *Erithacus rubecula*, Eurasian Blackbird *Turdus merula*, and Reed Bunting all showed elevated pectoralis muscle activities of CS and CCO in migratory relative to nonmigratory individuals (Lundgren and Kiessling 1985, 1986). Activities of CS and CCO in the pectoralis were also increased during migration in Goldcrests *Regulus regulus*, Great Tits *Parus major*, and Yellowhammers (Lundgren 1988) and pectoralis CS activity also increased during migration in Western Sandpipers (Guglielmo et al. 2002). These results suggest that alterations of mass-specific oxidative enzyme activities might function to elevate aerobic capacity in migrants or cold acclimatized birds.

In contrast to these data, however, are a number of studies documenting no mass-specific variation of oxidative enzyme activities with migration or cold acclimatization. Neither House Finches nor American Goldfinches exhibited differences in mass-specific CS activity of the pectoralis muscle associated with seasonal acclimatization, although leg muscle in House Finches did show a winter increment of CS activity (O'Connor 1995b; Carey et al. 1989; Yacoe and Dawson 1983; Marsh and Dawson 1982). CS activities in pectoralis, supracoracoideus, and leg muscles of White-breasted Nuthatches were also seasonally stable (Liknes 2005). Furthermore, American Goldfinches show no seasonal variation in the activity of succinate dehydrogenase, another Krebs cycle enzyme (Carey et al. 1978), and pectoralis muscle homogenates and isolated mitochondria show no seasonal variation in their capacity to oxidize fats or carbohydrates (Yacoe and Dawson 1983). Mass-specific succinate dehydrogenase activity in the pectoralis also did not vary between migratory and nonmigratory Rosy Pastors (George and Vallyathan 1964). Gray Catbirds *Dumetella carolinensis* demonstrated no variation in mass-specific CS or CCO activities in flight muscles between migratory and nonmigratory individuals (Marsh 1981). No differences in mass-specific pectoralis CS activity were evident between migratory and nonmigratory Sedge Warblers (Lundgren and Kiessling 1985). In addition, Semipalmated Sandpipers and Red Knots showed no migration-induced variation in mass-specific CS or CCO activities, respectively, in flight muscles (Weber and Piersma 1996; Driedzic et al. 1993). Thus, while cold, activity, or migration-induced increments of mitochondrial density or mass-specific oxidative enzyme activities do occur in some species, their occurrence is far from universal, and such changes do not appear to be required for migration or cold acclimatization.

It might also be expected that nutritional organs would show increases in metabolic intensity associated with migratory disposition or cold acclimatization to assist in support of the elevated energy demands of migration and thermoregulation in the cold. There are few data to address this possibility, but some studies have examined catabolic enzyme activities in the heart associated with migration and digestive enzyme activity and nutrient uptake capacities in the gut associated with cold acclimation. Stable or decreasing mass-specific aerobic capacity of the heart, as indicated by CS or CCO activities, with migratory disposition occurs in Semipalmated Sandpipers and Red Knots (Weber and Piersma 1996; Driedzic et al. 1993). Heart CS activity was also seasonally stable in European Starlings (Vezina and Williams 2005). Cold acclimation in Cedar Waxwings *Bombycilla cedrorum* produced no changes in digestive enzyme activities or nutrient uptake rates per unit intestine (McWilliams et al. 1999). Intestinal carrier-mediated glucose uptake was higher in cold-acclimated, exercised House Wrens than in warm-acclimated sedentary controls, but carrier-mediated glucose transport was an order of magnitude lower than passive glucose transport and proline-uptake rates in the gut did not vary between groups (Dykstra and Karasov 1992). Thus, the few available data suggest that mass-specific metabolic intensity of central organs does not vary substantially with migration or cold acclimatization. In contrast, Zheng et al. (2008) documented higher CCO activity in winter than in summer in liver of Eurasian Tree Sparrows.

3.4.4 Seasonal or Migration-Induced Changes in Organ Masses and Their Influence on Organismal Metabolic Rates

Because mass-specific increases in metabolic intensity do not seem to vary consistently with metabolic demands of migration or winter thermogenesis in birds, an alternative or additional strategy for elevating metabolic rates to meet these demands is to increase the masses of the tissues involved in provision and use of energy. A number of recent studies have examined the relationship between variation in masses of organs and variation in BMR in birds. The interspecific study of Daan et al. (1990) revealed that BMR in birds was correlated with lean dry masses of heart and kidney, but not with masses of other organs. BMR was also correlated with daily energy expenditure in this study, prompting the authors to conclude that organ mass is adjusted to meet daily energetic demands and that BMR varies accordingly with the adjusted organ masses. BMR varied positively with lean mass in Red Knots, but this relationship was mostly associated with variation in masses of stomach and intestine (Piersma et al. 1996a). In Tree Swallows *Tachycineta bicolor*, BMR was positively associated with kidney mass, but was negatively associated with pectoralis and intestine masses and showed no relationship with heart mass (Burness et al. 1998). Intestine length and liver mass were positively correlated with BMR in nestling European Shags *Phalacrocorax aristotelis*, but BMR showed no relationship with heart or kidney masses (Bech and Ostnes 1999). BMR was positively associated with gut, liver, kidney, and pectoralis masses in House Sparrows (Chappell et al. 1999). BMR was positively associated with intestine and lung masses in male Red Junglefowl and with spleen mass in females (Hammond et al. 2000). Cold-acclimated Hoopoe Larks *Alaemon alaudipes* showed elevated BMR relative to warm-acclimated birds and BMR was positively associated with masses of liver, kidney, intestine and stomach (Williams and Tieleman 2000). Vezina and Williams (2003) studied variation in BMR and organ masses during the breeding season in female European Starlings and found that BMR was positively associated with oviduct mass during the laying period, with liver and gizzard masses during the chick-rearing period, and with pectoral muscle mass during the nonbreeding (end of winter) season. In contrast, BMR decreased while gizzard mass increased in Red Knots shifted from soft to hard-shelled food (Piersma et al. 2004). These studies indicate that changes in organ masses can influence BMR in birds and although individual organs are not positively associated with BMR in every case, important contributors to BMR variation appear to include kidney, liver, heart, and gut masses.

Given that variation in masses of central organs can influence BMR in birds, it is instructive to examine how central organ masses vary with migratory disposition or with winter acclimatization. Increment of heart mass is a regular component of development of migratory disposition in birds (Piersma 1998). Similarly, gut mass increases are commonly associated with elevated energy expenditures and resultant increases in daily food intake in birds (McWilliams and Karasov 2001; Karasov 1996). For long-distance migratory shorebirds, and presumably for other long-distance migrant species as well, immediately prior to departure on migratory flights "exercise organs" (i.e., pectoralis muscle and heart) increase in mass, while "nutritional organs" (i.e., stomach, intestine, and liver) decrease in mass (Landys-Ciannelli et al. 2003; Battley et al. 2001; Piersma 1998). Instructive in this regard is a study of body composition variation during a 24-day migratory stopover in Red Knots on the southern coast of Iceland (Piersma et al. 1999). During the first 7-days of the stopover period, knots showed little change in overall body mass, although heart, stomach, and liver masses increased. Over the next 10-days, body mass increased rapidly as the birds accumulated fat stores and masses of kidneys, liver, and intestine increased, along with a decrease in stomach mass. During the final 7-days before departure, body mass increased slowly and pectoral muscle and heart masses also increased, while stomach, intestine, liver, and leg muscle masses decreased (Fig. 3.7). Similar patterns of organ mass variation have been documented for Great Knot *Calidris tenuirostris* (Battley et al. 2001) and Bar-tailed Godwit (Landys-Ciannelli et al. 2003).

Also interesting are examinations of the body composition of birds immediately before and/or after long-distance migratory flights. Biebach (1998) and Bauchinger et al. (2005) studied migratory Garden Warblers before and after crossing the Sahara desert (a 2–3-day flight period) and found that this crossing resulted in decreases in most organ masses, with breast and leg muscle masses decreasing from 14 to 25% and digestive organ masses decreasing from 34 to 57%. Lean dry masses of most organs (excepting brain and lung) decreased following a 5,400-km migratory flight from Australia to China in Great Knots (Battley et al. 2000). Similarly, individual Red Knots showed decreases in pectoral muscle size and body mass following extended flights in a laboratory wind tunnel (Lindström et al. 2000) and a Thrush Nightingale showed decreases of about 20% in both BMR and body mass following 12-h wind tunnel flights (Lindström et al. 1999).

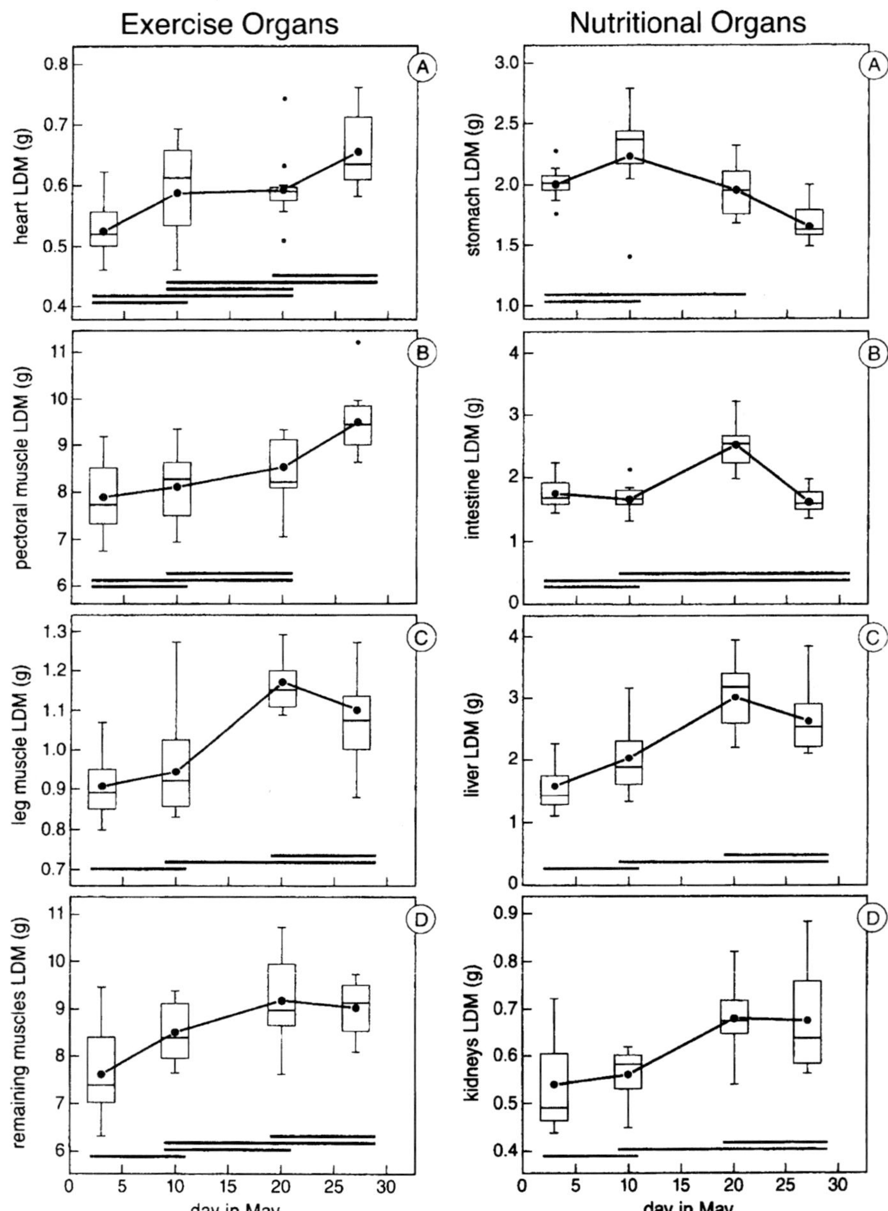

Fig. 3.7 Variation in lean dry mass of exercise (*left*) and nutritional (*right*) organs over a 24-day stopover during spring migration (May) in Red Knots, *Calidris canutus*. The *large dots* represent mean values, while the *horizontal lines* within the boxes represent median values. Boxes provide 25% quartiles around the median, while *error bars* indicate the range of values. Where *small dots* are present, they represent outlier values. *Heavy horizontal lines* indicate values not significantly different from each other. Reprinted from Piersma et al. (1999), with permission

Bar-tailed Godwits just prior to takeoff on long-distance migratory flights had small gizzard, gut, liver, and kidney masses relative to nonmigrating individuals (Piersma and Gill 1998). Migration in Western Sandpipers, a short-hop migrant, resulted in increases in masses of pectoralis muscle, heart, gizzard, pancreas, liver, and ceca, with relatively greater increases in digestive than in exercise organs (Guglielmo and Williams 2002).

Seasonal changes in gut mass, with higher values in winter, are common among birds, but these changes are usually associated with diet switches to less digestible food and gut mass increases on diets of low digestibility (Karasov 1990). However, there is evidence that cold, and its attendant thermoregulatory demands, can result in increments of gut mass in birds. Cold acclimation resulted in a 30% increase in intestinal length in Japanese Quail *Coturnix coturnix* (Fenna and Boag 1974). Dykstra and Karasov (1992) showed that cold acclimation in House Wrens produced a 21% increase in small intestine length. Moreover, cold-acclimated Cedar Waxwings increased masses of digestive organs by 22–53% over warm-acclimated controls (McWilliams et al. 1999). Winter acclimatized Rufous-collared Sparrows *Zonotrichia capensis* exhibited dry mass increases of 27 and 45% for crop and intestine, respectively, and this occurred despite the larger fraction of insects relative to seeds in the diets of winter birds relative to their summer counterparts (Novoa et al. 1996). Liknes (2005) documented winter increases, relative to summer, in intestine mass for Black-capped Chickadees, in liver and gizzard masses for House Sparrows, and in gizzard mass for White-breasted Nuthatches. Eurasian Tree Sparrows showed elevated BMR in winter along with increases in masses of liver, gizzard, and intestine (Zheng et al. 2008; Liu and Li 2006). Finally, Swanson (1991b) showed that liver wet mass in winter acclimatized Dark-eyed Juncos increased by 39% relative to summer acclimatized birds. Even though it is difficult to tease apart the factors driving the increase in gut and digestive organ masses during winter (i.e., diet or energetic demands of thermoregulation), elevated masses of the gut and digestive organs in winter may entail an energetic cost for maintenance, thereby driving up BMR in winter.

In summary, changes in central organ masses, and sometimes flight muscle masses, can influence BMR in birds and such changes are associated with migration and seasonal acclimatization. Thus, seasonal and migration-induced changes in body composition appear to be a mechanism by which similar variation in BMR can be modulated (e.g., Piersma 1998, 2002).

Can similar seasonal or migration-induced variation in organ masses help account for the seasonal patterns of variation in aerobic capacity or M_{sum} documented for birds? Changes in M_{sum} could presumably result from alterations in masses of skeletal muscles, especially those prominently involved in shivering (i.e., flight muscles). Few studies have directly examined correlations between aerobic capacity or M_{sum} and organ masses in birds. However, Chappell et al. (1999) investigated correlations between organ masses and maximum oxygen consumption during hop-flutter exercise in House Sparrows. In this study, breast muscle and heart masses were significantly and positively correlated with aerobic capacity, although they explained only 17% of the variation in aerobic capacity. Hammond et al. (2000) demonstrated that maximum oxygen consumption during treadmill running in Red Junglefowl was positively correlated with intestine mass in females and with heart and pectoralis mass in males. In addition, M_{sum} was positively correlated with breast muscle size in both cold-acclimated and migratory Red Knots (Vezina et al. 2006, 2007). These data suggest that masses of flight muscles and heart, among other factors, can have an important influence on aerobic capacity or M_{sum} in birds.

If we accept that variation in masses of skeletal muscle and heart can influence aerobic capacity and M_{sum} in birds, to determine if such changes can assist in explaining seasonal patterns of variation in M_{sum} we need to examine how muscle and heart masses change with migration and winter acclimatization. Flight muscle and heart hypertrophy are common in migratory birds during premigratory and migratory periods (Landys-Ciannelli et al. 2003; Piersma 1998; Dawson et al. 1983a). This hypertrophy may occur rapidly, over a period as short as a few days (Lindström et al. 2000; Piersma et al. 1999; Jehl 1997; Fry et al. 1972). Increases in pectoralis muscle mass associated with the development of migratory disposition can be as high as 35% (Piersma et al. 1999; Jehl 1997; Evans et al. 1992; Marsh 1984; Baggott 1975). The degree of pectoralis muscle hypertrophy can even exceed these levels in species rebuilding muscle following a flightless molt period prior to migration (Gaunt et al. 1990). Increments of heart mass are also commonly associated with preparedness for departure on migratory flights in birds (Landys-Ciannelli et al. 2003; Piersma 1998) and interspecific modeling studies have revealed that heart mass and associated stroke volume changes are important to adaptive specializations of aerobic capacity (Bishop and Butler 1995).

Pectoralis muscle hypertrophy is also a common aspect of winter acclimatization, at least in small birds with marked winter increases in M_{sum} (Liknes 2005; Cooper 2002; O'Connor 1995b; Swanson 1991b). These studies indicate that percent changes in pectoralis muscle mass often rather closely parallel percent changes in M_{sum} (Fig. 3.8). In contrast to these data, however, House Finches from Colorado showed no seasonal variation in pectoralis muscle mass, but this population also exhibited no seasonal difference in M_{sum} and only modest winter improvement of cold tolerance (Carey et al. 1989; Dawson et al. 1983b). Seasonal variability in heart mass associated with winter acclimatization in birds has received little study, but heart mass did increase in winter relative to summer for three species of small passerines (Black-capped Chickadees, White-breasted Nuthatches, and House Sparrows) from South Dakota (Liknes 2005).

In summary, available data indicate that flight muscle and/or heart do increase in mass with migratory disposition and winter acclimatization in birds. These adjustments likely impact aerobic capacity and M_{sum} during these periods of the annual cycle, as suggested by the data for winter acclimatized birds. In addition, because flight muscles comprise such a large fraction of total body mass in volant birds, hypertrophy of these muscles might also elevate BMR and such a correlation has been demonstrated in some birds (Vezina and Williams 2003; Chappell et al. 1999).

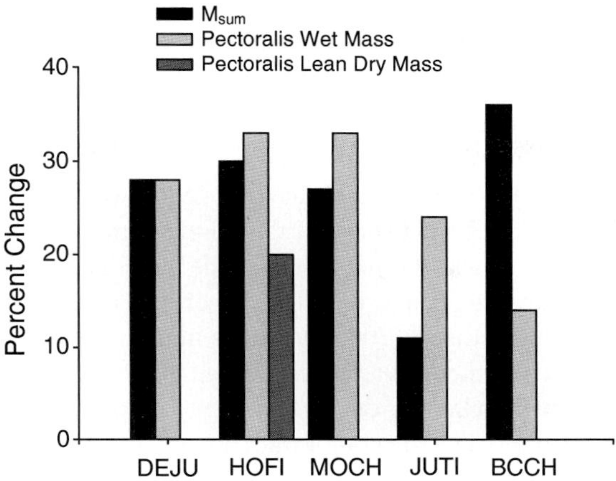

Fig. 3.8 Percent increases in pectoralis muscle mass and M_{sum} in winter relative to summer in passerine birds for which both values have been measured. Data are from the following studies: Dark-eyed Junco (DEJU, Swanson 1990a; 1991b), House Finch (HOFI, O'Connor 1995a, b), Mountain Chickadee and Juniper Titmouse (MOCH and JUTI, Cooper 2002), and Black-capped Chickadee (BCCH, Cooper and Swanson 1994, E.T. Liknes and Swanson, unpublished data). Pectoralis muscle mass values are mostly for wet mass because this is the most common measure reported in the literature, although lean dry mass is probably the better metric. O'Connor (1995b) provided seasonal measurements of pectoralis lean dry mass for House Finches, so I also included this value in the figure

3.5   Suggestions for Future Research

Seasonal patterns of variation in BMR and M_{sum} associated with winter acclimatization in birds have been well described. However, more research is needed to better describe patterns of variation in BMR and M_{sum} during the different stages of reproduction (e.g., Vezina and Williams 2005). Moreover, the functional significance of metabolic variation during reproduction and how such variation might be related to variation in FMR is in need of further study (e.g., Vezina and Williams 2003). BMR variation associated with migration has been well documented in shorebirds, but migration-associated variation in BMR is practically unstudied in other migratory birds. Further study is also required to document if emerging patterns of migration-induced M_{sum} variation are broadly applicable. Measurement of MMR variation associated with migration would also be fruitful, although current techniques measuring MMR during hovering flight or in hop-flutter wheels (Chappell et al. 1999; Chai and Dudley 1995, 1996) might not accurately represent MMR during horizontal flight typical of migration.

Several topics related to the relationship between organ masses, tissue metabolic intensity, and organismal metabolic rates in birds deserve further research attention. First, additional detailed experiments examining potential variation in both metabolic intensity (e.g., CS, CCO, HOAD activities, and/or mitochondrial density) and organ masses of nutritional and exercise organs, along with measurement of BMR and M_{sum}, or at least in species for which BMR and M_{sum} are known to vary with migration or season, are needed. To date, the only birds for which most of these parameters have been measured are Red Knots during migration (Piersma et al. 1995, 1996a, 1999; Weber and Piersma 1996) and American Goldfinches and House Finches during winter acclimatization (O'Connor 1995a, b; Carey et al. 1989; Dawson et al. 1983b; Yacoe and Dawson 1983; Marsh and Dawson 1982; Dawson and Carey 1976). Recent studies examining organ mass changes and metabolic intensity of European Starlings during the breeding season (Vezina and Williams 2005) and several species of small passerines during seasonal acclimatization (Zheng et al. 2008; Liknes 2005) are steps in the right direction.

Second, virtually all studies of variation in organ masses have so far been carried out on samples from populations during different seasons or different stages of migration, rather than the ideal method of tracking organ masses in individual birds over time. Following temporal variation in organ masses in individual birds has not been possible because, as Lindström and Piersma (1993, p. 70) point out, "one cannot kill a bird twice" for examination of body composition on individual birds during periods of changing energy demands. However, Dietz et al. (1999a) used ultrasonography to demonstrate that organ masses in individual captive Red Knots did change over time in association with the development of migratory disposition. Lindström et al. (2000) also used ultrasonography to document changes in pectoral muscle size associated with flight, fasting, and fueling in individual knots. To date, ultrasonography has been validated for measurement of breast muscle size in birds as small as medium-sized shorebirds, but it is likely applicable to smaller species as well. Thus, recent technological developments, including ultrasonography (Lindström et al. 2000; Dietz et al. 1999a, 1999b) and high field MRI (Anderson et al. 2000), may allow successful noninvasive tracking of organ size changes in individual birds over time. Studies using such techniques during the development of migratory disposition or during exposure to cold temperatures would nicely complement current data and coupling organ size data with metabolic measurements would allow assessment of how tightly changes in organ masses and metabolism are linked in individual birds. Indeed, Vezina et al. (2006, 2007) tracked changes in breast muscle size during cold acclimation and development of migratory disposition using ultrasonography and correlated these changes with BMR and M_{sum}. In addition, such studies could show how rapidly mass and metabolism could be adjusted to meet changing energetic demands.

A third area where additional research is needed is to determine if body composition or metabolic intensity alterations confer differences in endurance flight or cold hardiness. Current data suggesting such changes are largely correlative. Experiments monitoring cold or exercise training effects on flight or shivering endurance, BMR, M_{sum}, and MMR in association with measurements of organ mass and metabolic intensity variation would be useful in this regard. Additionally, the question of whether exercise training influences thermogenic capacity or endurance, or whether cold acclimation increases maximum exercise-induced metabolism or exercise endurance has not been addressed in birds. Potentially relevant in this regard is the finding that M_{sum} and MMR during activity were significantly and positively associated in deer mice *Peromyscus maniculatus* (Hayes and Chappell 1990). Cold acclimation increases peak metabolism during exercise in some mammals (Turner et al. 1995; Hayes and Chappell 1986) and exercise training elevates M_{sum} or cold tolerance in others (McDonald et al. 1988; Harri et al. 1984; Strømme and Hammel 1967). However, in some mammals exercise training does not influence M_{sum} (Conley et al. 1985). Moreover, extrapolation of mammalian results to birds is complicated by the prominent role of brown fat and NST in mammalian thermoregulation in the cold. The only relevant study for birds of which I am aware is Vezina et al. (2007), which documented that increases in body mass and flight muscle size associated with migratory disposition in Red Knots were positively correlated with M_{sum}.

Because fat is the principal fuel for both long-distance migratory flights and prolonged shivering, additional studies of what factors might limit fat catabolism, and thereby influence organismal metabolic rates, during migratory flights and shivering are needed. Recent studies suggest that intracellular transport or transport across sarcolemmal and mitochondrial membranes may be critical steps in regulating lipid catabolism (McWilliams et al. 2004). Research addressing variation in these steps with migration is in its infancy (McFarlan 2007; Guglielmo et al. 2002; Pelsers et al. 1999) and even less is known about variation in these steps with seasonal acclimatization (Liknes 2005). Another factor relating to fat catabolism and organismal metabolic rates that warrants further study is the impact of fatty acid composition of depot fat on aerobic performance in birds. Pierce et al. (2005) studied Red-eyed Vireos and showed that diets lower in unsaturated fatty acids (but higher in 18:2n6 content) produced fat stores with similar fatty acid composition, and that birds with lower unsaturated fatty acid content (but higher 18:2n6 content) had elevated MMR during hop-flutter wheel exercise. How such differences in fatty acid composition are regulated naturally in birds and how such differences might impact migratory performance or M_{sum} would be a fruitful area for additional research.

The cellular and molecular mechanisms regulating muscle and organ mass changes with migration and seasonal acclimatization have received very little study to date. A potential candidate for regulation of muscle mass changes is myostatin, a member of the TGF-β superfamily of growth factors, which is a potent autocrine/paracrine inhibitor of muscle growth in mammals (Lee 2004) and birds (Kim et al. 2006, 2007). Myostatin is synthesized in skeletal muscle in an inactive form that requires proteolytic removal of the N-terminal signal sequence and the propeptide to produce the active C-terminal fragment (Lee and McPherron 2001; McPherron and Lee 1997). Cleavage of the latent complexes to the mature form that binds to myostatin receptors is required for myostatin activity. Metalloproteinases, including BMP-1/tolloid family members TLL-1 and TLL-2, can activate myostatin (Wolfman et al. 2003; Huet et al. 2001). Swanson et al. (2009) documented winter decreases in myostatin and TLL-1 gene expression in pectoralis muscle of House Sparrows, a species showing winter increases in pectoralis muscle mass. These results are consistent with a role for myostatin in promoting pectoralis muscle hypertrophy in winter and additional research examining how expression and protein levels of myostatin and its metalloproteinase activators vary with migration and seasonal acclimatization would likely be productive.

Finally, one proposal for what limits maximal thermogenic capacity in birds is the amount of muscle recruited (Marsh and Dawson 1989). Because birds shiver isometrically (Hohtola 1982), force production by the smaller of the antagonistic muscle pair (supracoracoideus in the case of flight muscles) may limit force production by the larger muscle. Thus, for winter acclimatized birds, measurement of relative seasonal changes in mass and metabolic intensity of pectoralis and supracoracoideus muscles could illuminate mechanisms for enhancing thermogenic capacity and cold hardiness. Relatively larger seasonal changes in the supracoracoideus muscle might be expected if the smaller of the muscle pair is actually a limiting factor to thermogenic capacity. Such a scenario would not be expected in migrating birds because pectoralis and supracoracoideus muscles contract sequentially rather than isometrically during flight. Indeed, in Gray Catbirds, mass-specific cytochrome c concentrations and CS activities in the pectoralis were approximately double values for the supracoracoideus, although supracoracoideus CS activity was weakly, but positively, correlated with premigratory fattening (Marsh 1981). However, the relative masses of pectoralis and supracoracoideus muscles vary during the annual cycle in Red Knots, although such variation is not necessarily associated with migration (Piersma and Dietz 2007). Measurements of relative masses and metabolic intensities of pectoralis and supracoracoideus muscles during winter acclimatization and migration are needed.

3.6 Summary and Conclusions

In this paper, I have reviewed the evidence for seasonal and migration-induced variation in metabolic rates in birds. I have also tried to draw some generalizations about the importance of variation in BMR and M_{sum} to the ecology of wintering and migrating birds and attempted to link seasonal or migration-induced variation in organismal metabolic rates with physiological and biochemical variation at the cellular and tissue levels. Several conclusions that I believe are noteworthy have emerged from this body of work.

First, increments of M_{sum} in winter acclimatized small birds relative to their summer counterparts appear to be greater than previously recognized, and such increments are a common component of winter acclimatization in these birds. Thus, it seems probable that the mechanistic adjustments underlying winter increases in shivering endurance (i.e., cold tolerance) also promote increases in M_{sum}. Elevated M_{sum} in winter should increase heat production for any given fraction of M_{sum} that is sustained during shivering, and this would appear to be the functionally significant aspect of winter increases in M_{sum} to improved cold hardiness. However, variation in BMR does not necessarily track changes in cold tolerance and M_{sum} in birds, so its precise relation to winter improvements in cold resistance is uncertain and may vary among species.

Second, BMR and M_{sum} (and presumably maximal exercise-induced metabolic rates) also vary with migratory disposition in birds, although spring and fall migratory periods do not necessarily induce identical changes in organismal metabolism. From the limited available data, migration-induced changes in M_{sum} appear better explained as a by-product of adjustments for endurance flight, rather than for adjustment to cold temperatures. Nevertheless, higher M_{sum} in spring migrants may benefit birds if they encounter cold or adverse weather during migration or during arrival on the breeding grounds. Migration-induced changes in BMR appear, in general, to track changes in energy demand during the migratory period.

Third, mechanistic changes underlying variation in metabolic rates in birds could occur at several levels of the interacting system for energy processing, transport, and use. Some levels of this interacting system have received far more study than others, but some preliminary conclusions have emerged from this body of work. The ability to transport oxygen in the blood and to deliver it to working muscles seems to parallel energy demands from the tissues in species studied to date. Lipid transport and delivery to muscle cells does not appear to vary consistently with season or migratory disposition. Intracellular transport of lipid to the mitochondria, as well as intracellular stores of lipid, do seem positively related to energy demands, but this topic has been very little studied. Mass-specific aerobic capacity of muscles does not vary consistently with migration or winter acclimatization, nor does the metabolic intensity of central organs. Organ mass changes, however, do occur regularly with migration and winter acclimatization and this variation appears to occur in an adaptive manner, with larger central organ and muscle masses as needs for energy processing and/or use increase. Such changes in organ masses appear to serve as a vehicle by which whole-organism changes in aerobic capacity and metabolic rates are affected.

In summary, some aspects relating to variation in metabolic rates in birds and their underlying mechanisms have been well studied, but other aspects have received little attention. Thus, some of the preceding generalizations must be regarded as preliminary. More research needs to be conducted to thoroughly evaluate these generalizations and I have tried to outline some potentially valuable research approaches toward this end throughout the review. Data from such studies should help solidify our understanding of how seasonal metabolic variation in birds is mediated and whether or not such variation occurs in an adaptive manner.

Acknowledgments I thank Eric Liknes, Theunis Piersma, and an anonymous reviewer for comments that improved an earlier version of this manuscript. I also thank Eric Liknes for help with constructing some of the figures.

References

Abumrad N, Coburn C, Ibrahimi A (1999) Membrane proteins implicated in long-chain fatty acid uptake by mammalian cells: CD36, FATP and FABPm. Biochim Biophys Acta 1441:4–13

Ambrose SJ, Bradshaw SD (1988) Seasonal changes in standard metabolic rates in the White-browed Scrubwren *Sericornis frontalis* (Acanthizidae) from arid, semi-arid, and mesic environments. Comp Biochem Physiol 89A:79–83

Anava A, Kam M, Shkolnik A, Degen AA (2000) Seasonal field metabolic rate and dietary intake in Arabian Babblers (*Turdoides squamiceps*) inhabiting extreme deserts. Functional Ecol 14:607–613

Anava A, Kam M, Shkolnik A, Degen AA (2002) Seasonal daily, daytime and night-time field metabolic rates in Arabian Babblers. J Exp Biol 205:3571–3575

Anderson CL, Kabalka GW, Layne DG, Dyke JP, Burghardt GM (2000) Noninvasive high field MRI brain imaging of the Garter Snake (*Thamnophis sirtalis*). Copeia 2000:265–269

Arens JR, Cooper SJ (2005a) Metabolic and ventilatory acclimatization to cold stress in House Sparrows (*Passer domesticus*). Physiol Biochem Zool 78:579–589

Arens JR, Cooper SJ (2005b) Seasonal and diurnal variation in metabolism and ventilation in House Sparrows. Condor 107:433–444

Aulie A, Tøien O (1988) Threshold for shivering in aerobic and anaerobic muscles in bantam cocks and incubating hens. J Comp Physiol B 158:431–435

Baggott GK (1975) Moult, flight muscle "hypertrophy" and premigratory lipid deposition of the juvenile Willow Warbler, *Phylloscopus trochilus*. J Zool Lond 175:299–314

Barré H, Bailly L, Rouanet JL (1987) Increased oxidative capacity in skeletal muscles from cold acclimated ducklings: a comparison with rats. Comp Biochem Physiol 88B:519–522

Battley PF, Piersma T, Dietz MW, Tang S, Dekinga A, Hulsman K (2000) Empirical evidence for differential organ reductions during trans-oceanic bird flight. Proc R Soc Lond B 267:191–195

Battley PF, Dietz MW, Piersma T, Dekinga A, Tang S, Julsman K (2001) Is long-distance bird flight equivalent to a high-energy fast? Body composition changes in freely migrating and captive fasting Great Knots. Physiol Biochem Zool 74:435–449

Bauchinger U, Wohlmann A, Biebach H (2005) Flexible remodeling of organ size during spring migration of the Garden Warbler (*Sylvia borin*). Zoology 108:97–106

Bech C, Ostnes JE (1999) Influence of body composition on the metabolic rate of nestling European Shags (*Phalacrocorax aristotelis*). J Comp Physiol B 169:263–270

Bech C, Præstang KE (2004) Thermoregulatory use of heat increment of feeding in the Tawny Owl (*Strix aluco*). J Therm Biol 29:649–654

Bennett AF (1991) The evolution of activity capacity. J Exp Biol 160:1–23

Bennett PM, Harvey PH (1987) Active and resting metabolism in birds: allometry, phylogeny and ecology. J Zool Lond 213:327–363

Biebach H (1998) Phenotypic organ flexibility in Garden Warblers *Sylvia borin* during long-distance migration. J Avian Biol 29:529–535

Bishop CM, Butler PJ (1995) Physiological modelling of oxygen consumption in birds during flight. J Exp Biol 198:2153–2163

Blem CR, Blem LB (2006) Variation in mass of female Prothonotary Warblers during nesting. Wilson J Ornithol 118:3–12

Bonen A, Campbell SE, Benton CR, Chabowski A, Coort SLM, Han X-X, Koonen DPY, Glatz JFC, Luiken JJFP (2004) Regulation of fatty acid transport by fatty acid translocase/CD36, Proc. Nutr Soc 63:245–249

Bordel R, Haase E (1993) Effects of flight on blood parameters in homing pigeons. J Comp Physiol B 163:219–224

Brackenbury J (1984) Physiological responses of birds to flight and running. Biol Rev 59:559–575

Breuer K, Lill A, Baldwin J (1995) Haematological and body-mass changes of small passerines overwintering in south-eastern Australia. Aust J Zool 43:31–38

Brigham RM (1992) Daily torpor in a free-ranging goatsucker, the Common Poorwill (*Phalaenoptilus nuttallii*). Physiol Zool 65:457–472

Brigham RM, Trayhurn P (1994) Brown fat in birds? A test for the mammalian BAT-specific mitochondrial uncoupling protein in Common Poorwills. Condor 96:208–211

Brigham RM, Woods CP, Lane JE, Fletcher QE, Geiser F (2006) Ecological correlates of torpor use among five caprimulgiform birds. Acta Zoologica Sinica 52 (suppl):401–404

Broggi J, Hohtola E, Koivula K, Orell M, Thomson RL, Nilsson J-A (2007) Sources of variation in winter basal metabolic rate in the great tit. Funct Ecol 21:528–533

Bruinzeel LW, Piersma T (1998) Cost reduction in the cold: heat generated by terrestrial locomotion partly substitutes for thermoregulation costs in Knot *Calidris canutus*. Ibis 140:323–328

Bryant DM (1988) Energy expenditure and body mass changes as measures of reproductive costs in birds. Funct Ecol 2:23–34

Bundle MW, Hoppeler H, Vock R, Tester JM, Weyand PG (1999) High metabolic rates in running birds. Nature 397:31–32

Burness GP, Ydenberg RC, Hochachka PW (1998) Interindividual variability in body composition and resting oxygen consumption rate in breeding Tree Swallows, *Tachycineta bicolor*. Physiol Zool 71:247–256

Butler PJ, Turner DL (1988) Effect of training on maximal oxygen uptake and aerobic capacity of locomotory muscles in Tufted Ducks, *Aythya fuligula*. J Physiol 401:347–359

Campbell SE, Tandon NN, Woldegiorgis G, Luiken JJFP, Glatz JFC, Bonen A (2004) A novel function for fatty acid translocase (FAT)/CD36. J Biol Chem 35:36235–36241

Carey C, Dawson WR, Maxwell LC, Faulkner JA (1978) Seasonal acclimatization to temperature in cardueline finches. II. Changes in body composition and mass in relation to season and acute cold stress. J Comp Physiol 125:101–113

Carey C, Marsh RL, Bekoff A, Johnston RM, Olin AM (1989) Enzyme activities in muscles of seasonally acclimatized House Finches. In: Bech C, Reinertsen RE (eds) Physiology of cold adaptation in birds. Plenum Life Sciences, New York, pp 95–104

Cavieres G, Sabat P (2008) Geographic variation in the response to thermal acclimation in Rufous-collared Sparrows: are physiological flexibility and environmental heterogeneity correlated? Funct Ecol 22:509–515

Chai P, Dudley R (1995) Limits to vertebrate locomotor energetics suggested by hummingbirds hovering in heliox. Nature 377:722–725

Chai P, Dudley R (1996) Limits to flight energetics of hummingbirds hovering in hypodense and hypoxic gas mixtures. J Exp Biol 199:2285–2295

Chai P, Chang AC, Dudley R (1998) Flight thermogenesis and energy conservation in hovering hummingbirds. J Exp Biol 201:963–968

Chappell MA, Bachman GC, Hammond KA (1997) The heat increment of feeding in House Wren chicks: magnitude, duration, and substitution for thermostatic costs. J Comp Physiol B 167:313–318

Chappell MA, Bech C, Buttemer WA (1999) The relationship of central and peripheral organ masses to aerobic performance variation in House Sparrows. J Exp Biol 202:2269–2279

Chastel O, Kersten M (2002) Brood size and body condition in the House Sparrow *Passer domesticus*: the influence of brooding behavior. Ibis 144:284–292

Cichon M (2001) Body-mass changes in female Collared Flycatchers: state-dependent strategy. Auk 118:550–552

Collin A, Buyse J, Van As P, Darras VM, Malheiros RD, Moraes VMB, Reyns GE, Taouis M, Decuypere E (2003a) Cold-induced enhancement of avian uncoupling protein expression, heat production, and triiodothyronine concentrations in broiler chicks. Gen Comp Endocrinol 130:70–77

Collin A, Taouis M, Buyse J, Ifuta NB, Darras VM, Van As P, Malheiros RD, Moraes VMB, Decuypere E (2003b) Thyroid status, but not insulin status, affects expression of avian uncoupling protein mRNA in chicken. Am J Physiol Endocrinol Metab 284:E771–E777

Conley KE, Weibel ER, Taylor CR, Hoppeler H (1985) Aerobic capacity estimated by exercise vs cold exposure: endurance training effects in rats. Respir Physiol 62:273–280

Cooper SJ (2000) Seasonal energetics of Mountain Chickadees and Juniper Titmice. Condor 102:635–644

Cooper SJ (2002) Seasonal metabolic acclimatization in Mountain Chickadees and Juniper Titmice. Physiol Biochem Zool 74:386–395

Cooper SJ, Same DR (2000) Ventilatory accommodation under cold stress in seasonally acclimatized Black-capped Chickadees. Am Zool 40:980A

Cooper SJ, Sonsthagen S (2007) Heat production from foraging activity contributes to thermoregulation in Black-capped Chickadees. Condor 109:446–451

Cooper SJ, Swanson DL (1994) Seasonal acclimatization of thermoregulation in the Black-capped Chickadee. Condor 96:638–646

Cottam M, Houston D, Lobley G, Hamilton I (2002) The use of muscle protein for egg production in the Zebra Finch *Taeniopygia guttata*. Ibis 144:210–217

Croll DA, Gaston AJ, Noble DG (1991) Adaptive loss of mass in Thick-billed Murres. Condor 93:496–502

Daan S, Masman D, Groenewold A (1990) Avian basal metabolic rates: their association with body composition and energy expenditure in nature. Am J Physiol 259:R333–R340

Dawson WR (1958) Relation of oxygen consumption and evaporative water loss to temperature in the cardinal. Physiol Zool 31:37–48

Dawson WR, Carey C (1976) Seasonal acclimatization to temperature in cardueline finches. I. insulative and metabolic adjustments. J Comp Physiol 112:317–333

Dawson WR, Marsh RL (1989) Metabolic acclimatization to cold and season in birds. In: Bech C, Reinertsen RE (eds) Physiology of cold adaptation in birds. Plenum Life Sciences, New York, pp 83–94

Dawson WR, O'Connor TP (1996) Energetic features of avian thermoregulatory responses. In: Carey C (ed) Avian energetics and nutritional ecology. Chapman & Hall, New York, pp 85–124

Dawson WR, Smith BK (1986) Metabolic acclimatization in the American Goldfinch (*Carduelis tristis*). In: Heller HC, Musacchia XJ, Wang LCH (eds) Living in the cold: physiological and biochemical adaptations. Elsevier, New York, pp 427–437

Dawson WR, Marsh RL, Yacoe ME (1983a) Metabolic adjustments of small passerine birds for migration and cold. Am J Physiol 245:R755–R767

Dawson WR, Marsh RL, Buttemer WA, Carey C (1983b) Seasonal and geographic variation of cold resistance in house finches. Physiol Zool 56:353–369

deGraw WA, Kern MD, King JR (1979) Seasonal changes in the blood composition of captive and free-living White-crowned Sparrows. J Comp Physiol B 129:151–162

Dietz MW, Piersma T, Dekinga A (1999a) Body-building without power training: endogenously regulated pectoral muscle hypertrophy in confined shorebirds. J Exp Biol 202:2831–2837

Dietz MW, Dekinga A, Piersma T, Verhulst S (1999b) Estimating organ size in small migrating shorebirds with ultrasonography: an intercalibration exercise. Physiol Biochem Zool 72:28–37

Doherty PF Jr, Williams JB, Grubb TC Jr (2001) Field metabolism and water flux of Carolina Chickadees during breeding and nonbreeding seasons: a test of the "peak-demand" and "reallocation" hypotheses. Condor 103:370–375

Driedzic WR, Crowe HL, Hicklin PW, Sephton DH (1993) Adaptations of pectoralis muscle, heart mass, and energy metabolism during premigratory fattening in Semipalmated Sandpipers (*Calidris pusilla*). Can J Zool 71:1602–1608

Duchamp C, Barre H (1993) Skeletal muscle as the major site of nonshivering thermogenesis in cold-acclimated ducklings. Am J Physiol 265:R1076–R1083

Duchamp C, Cohen-Adad F, Rouanet J-L, Dumonteil E, Barre H (1993) Existence of nonshivering thermogenesis in birds. In: Carey C, Florant GL, Wunder BA, Horwitz B (eds) Life in the cold: ecological, physiological, and molecular mechanisms. Westview Press, Boulder, Colorado, pp 529–533

Dutenhoffer MS, Swanson DL (1996) Relationship of basal to summit metabolic rate in passerine birds and the aerobic capacity model for the evolution of endothermy. Physiol Zool 69:1232–1254

Dykstra CR, Karasov WH (1992) Changes in gut structure and function of House Wrens (*Troglodytes aedon*) in response to increased energy demands. Physiol Zool 65:422–442

Ellegren H (1993) Speed of migration and migratory flight lengths of passerine birds ringed during autumn migration in Sweden. Ornis Scand 24:220–228

Ellis HI (1984) Energetics of free-ranging seabirds. In: Whittow GC, Rahn H (eds) Seabird energetics. Plenum, New York, pp 203–234

Emre Y, Hurtaud C, Ricquier D, Bouillaud F, Hughes J, Criscuolo F (2007) Avian UCP: the killjoy in the evolution of the mitochondrial uncoupling proteins. J Mol Evol 65:392–402

Evans PR, Davidson NC, Uttley JD, Evans RD (1992) Premigratory hypertrophy of flight muscles: an ultrastructural study. Ornis Scand 23:238–243

Fair J, Whitaker S, Pearson B (2007) Sources of variation in haematocrit in birds. Ibis 149:535–552

Fenna L, Boag DA (1974) Adaptive significance of the caeca in Japanese Quail and Spruce Grouse (Galliformes). Can J Zool 52:1577–1584

Fransson T (1995) Timing and speed of migration in North and West European populations of *Sylvia* warblers. J Avian Biol 26:39–48

Freed LM (1981) Loss of mass in breeding wrens: stress or adaptation? Ecology 62:1179–1186

French AR (1993) Hibernation in birds: comparisons with mammals. In: Carey C, Florant GL, Wunder BA, Horwitz B (eds) Life in the cold: ecological, physiological, and molecular mechanisms. Westview Press, Boulder, Colorado, pp 43–53

Fry CH, Ferguson-Lees IJ, Dowsett RJ (1972) Flight muscle hypertrophy and ecophysiological variation of Yellow Wagtail *Motacilla flava* races at Lake Chad. J Zool 167:293–306

Fyhn M, Gabrielsen GW, Nordøy ES, Moe B, Langseth I, Bech C (2001) Individual variation in field metabolic rate of kittiwakes (*Rissa tridactyla*) during the chick-rearing period. Physiol Biochem Zool 74:343–355

Gannes LZ, Hatch KA, Pinshow B (2001) How does time since feeding affect the fuels pigeons use during flight? Physiol Biochem Zool 74:1–10

Gaunt AS, Hikida RS, Jehl JR Jr, Fenbert L (1990) Rapid atrophy and hypertrophy of an avian flight muscle. Auk 107:649–659

Gelineo S (1964) Organ systems in adaptation: the temperature regulating system. In: Dill DB (ed) Handbook of physiology, Section 4, adaptation to the environment. American Physiological Society, Washington, D.C., pp 259–282

George JC, Vallyathan NV (1964) Lipase and succinic dehydrogenase activity of the particulate fractions of the breast muscle homogenate of the migratory starling *Sturnus roseus* in the pre-migratory and post-migratory periods. Can J Physiol Pharmacol 42:447–452

Gessaman JA, Nagy KA (1988) Transmitter loads affect the flight speed and metabolism of homing pigeons. Condor 90:662–668

Gessaman JA, Workman GW, Fuller MR (1991) Flight performance, energetics and water turnover of Tippler Pigeons with a harness and dorsal load. Condor 93:546–554

Goldstein DL (1990) Energetics of activity and free living in birds. Stud Avian Biol 13:423–426

Golet GH, Irons DB (1999) Raising young reduces body condition and fat stores in Black-legged Kittiwakes. Oecologia 120:530–538

Guglielmo CG, Williams TD (2002) Phenotypic flexibility of body composition in relation to migratory state, age and sex in the Western Sandpiper (*Calidris mauri*). Physiol Biochem Zool 76:84–98

Guglielmo CG, Haunerland NH, Williams TD (1998) Fatty acid binding protein, a major protein in the flight muscle of migrating Western Sandpipers. Comp Biochem Physiol B 119:549–555

Guglielmo CG, Haunerland NH, Hochachka PW, Williams TD (2002) Seasonal dynamics of flight muscle fatty acid binding protein and catabolic enzymes in a migratory shorebird. Am J Physiol 282:R1405–R1413

Gwinner E (ed) (1990) Bird migration: physiology and ecophysiology. Springer, Berlin

Hagan JM, Lloyd-Evans TL, Atwood JL (1991) The relationship between latitude and the timing of spring migration of North American landbirds. Ornis Scand 22:129–136

Hails CJ (1983) The metabolic rate of tropical birds. Condor 85:61–65

Hammond KA, Chappell MA, Cardullo RA, Lin R-I, Johnsen TS (2000) The mechanistic basis of aerobic performance variation in red junglefowl. J Exp Biol 203:2053–2064

Harms SJ, Hickson RC (1983) Skeletal muscle mitochondria and myoglobin, endurance, and intensity of training. J Appl Physiol 54:798–802

Harri M, Dannenberg T, Oksanen-Rossi R, Hohtola E, Sundin U (1984) Related and unrelated changes in response to exercise and cold in rats: a reevaluation. J Appl Physiol 57:1489–1497

Hart JS (1962) Seasonal acclimatization in four species of small wild birds. Physiol Zool 35:224–236

Hartman FA (1961) Locomotor mechanisms of birds. Smithsonian Misc Coll 143:1–91

Hawkins PAJ, Butler PJ, Woakes AJ, Gabrielsen GW (1997) Heat increment of feeding in Brunnich's Guillemot, *Uria lomvia*. J Exp Biol 200:1757–1763

Hayes JP, Chappell MA (1986) Effects of cold acclimation on maximum oxygen consumption during cold exposure and treadmill exercise in deer mice, *Peromyscus maniculatus*. Physiol Zool 59:473–481

Hayes JP, Chappell MA (1990) Individual consistency of maximal oxygen consumption in deer mice. Funct Ecol 4:495–503

Hickson RC (1981) Skeletal muscle cytochrome c and myoglobin, endurance, and frequency of training. J Appl Physiol 51:746–749

Hohtola E (1982) Thermal and electromyographic correlates of shivering thermogenesis in the pigeon. Comp Biochem Physiol 73A:159–166

Hohtola E (2002) Facultative and obligatory thermogenesis in young birds: a cautionary note. Comp Biochem Physiol A 131:733–739

Hohtola E, Henderson RP, Rashotte ME (1998) Shivering thermogenesis in the pigeon: the effects of activity, diurnal factors, and feeding state. Am J Physiol 275:R1553–R1562

Holloway GP, Bezaire V, Heigenhauser GJF, Tandon NN, Glatz JFC, Luiken JJFP, Bonen A, Spriet LL (2006) Mitochondrial long chain fatty acid oxidation, fatty acid translocase/CD36 content and carnitine palmitoyltransferase 1 activity in human skeletal muscle during aerobic exercise. J Physiol 571(1):201–210

Holt S, Whitfield DP, Duncan K, Rae S, Smith RD (2002) Mass loss in incubating Eurasian Dotterel: adaptation or constraint? J Avian Biol 33:219–224

Hoppeler H, Weibel ER (1998) Limits for oxygen and substrate transport in mammals. J Exp Biol 201:1051–1064

Houston DC, Donnan D, Jones P, Hamilton I, Osborne D (1995) Changes in the muscle condition of female Zebra Finches *Poephila guttata* during egg laying and the role of protein storage in bird skeletal muscle. Ibis 137:322–328

Huet C, Li Z-F, Liu H-Z, Black RA, Galliano M-F, Engvall E (2001) Skeletal muscle cell hypertrophy induced by inhibitors of metalloproteinases; myostatin as a potential mediator. Am J Physiol 281:C1624–C1634

Jaeger EC (1949) Further observations on the hibernation of the Poor-will. Condor 51:105–109

Jehl JR Jr (1994) Field estimates of energetics in migrating and downed Black-necked Grebes. J Avian Biol 25:63–68

Jehl JR Jr (1997) Cyclical changes in body composition in the annual cycle and migration of the Eared Grebe *Podiceps nigricollis*. J Avian Biol 28:132–142

Jenni-Eiermann S, Jenni L (1992) High plasma triglyceride levels in small birds during migratory flight: a new pathway for fuel supply during endurance locomotion at very high mass-specific metabolic rates. Physiol Zool 65:112–123

Jenni-Eiermann S, Jenni L (2001) Postexercise ketosis in night-migrating passerine birds. Physiol Biochem Zool 74:90–101

Jenni-Eiermann S, Jenni L, Kvist A, Lindström Å, Piersma T, Visser GH (2002) Fuel use and metabolic response to endurance exercise: a wind tunnel study of a long-distance migrant shorebird. J Exp Biol 205:2453–2460

Jones MM (1991) Muscle protein loss in laying House Sparrows *Passer domesticus*. Ibis 133:193–198

Jones IL (1994) Mass changes of Least Auklets *Aethia pusilla* during the breeding season: evidence for programmed loss of mass. J Anim Ecol 63:71–78

Karasov WH (1990) Digestion in birds: chemical and physiological determinants and ecological implications. Stud Avian Biol 13:391–415

Karasov WH (1996) Digestive plasticity in avian energetics and feeding ecology. In: Carey C (ed) Avian energetics and nutritional ecology. Chapman & Hall, New York, pp 61–84

Kaseloo PA, Lovvorn JR (2003) Heat increment of feeding and thermal substitution in mallard ducks feeding voluntarily on grain. J Comp Physiol B 173:207–213

Kersten M, Piersma T (1987) High levels of energy expenditure in shorebirds; metabolic adaptations to an energetically expensive way of life. Ardea 75:175–187

Kersten M, Bruinzeel LW, Wiersma P, Piersma T (1998) Reduced basal metabolic rate of migratory waders wintering in coastal Africa. Ardea 86:71–80

Kim YS, Bobbili NK, Paek KS, Jin HJ (2006) Production of a monoclonal anti-myostatin antibody and the effects of in ovo administration of the antibody on posthatch broiler growth and muscle mass. Poult Sci 85:1062–1071

Kim YS, Bobbili NK, Lee YK, Jin HJ, Dunn MA (2007) Production of a polyclonal anti-myostatin antibody and the effects of in ovo administration of the antibody on posthatch broiler growth and muscle mass. Poult Sci 86:1196–1205

Klaassen M (1995) Moult and basal metabolic costs in males of two species of stonechats: the European *Saxicola torquata rubicula* and the East African *S. t. axillaris*. Oecologia 104:424–432

Klaassen M (1996) Metabolic constraints on long-distance migration. J Exp Biol 199:57–64

Klaassen M, Biebach H (1994) Energetics of fattening and starvation in the long-distance migratory garden warbler, *Sylvia borin*, during the migratory phase. J Comp Physiol B 164:362–371

Klaassen M, Lindström Å, Zijlstra R (1997) Composition of fuel stores and digestive limitations to fuel deposition rate in the long-distance migratory thrush nightingale, *Luscinia luscinia*. Physiol Zool 70:125–133

Klaassen M, Kvist A, Lindström Å (2000) Flight costs and fuel composition of a bird migrating in a wind tunnel. Condor 102:444–451

Kullberg C, Metcalfe NB, Houston DC (2002) Impaired flight ability during incubation in the Pied Flycatcher. J Avian Biol 33:179–183

Landys MM, Piersma T, Guglielmo CG, Jukema J, Ramenofsky M, Wingfield JC (2005) Metabolic profile of long-distance migratory flight and stopover in a shorebird. Proc Roy Soc B 272:295–302

Landys-Ciannelli MM, Jukema J, Piersma T (2002) Blood parameter changes during stopover in a long-distance migratory shorebird, the Bar-tailed Godwit *Limosa lapponica taymyrensis*. J Avian Biol 33:451–455

Landys-Ciannelli MM, Piersma T, Jukema J (2003) Strategic size changes of internal organs and muscle tissue in the Bar-tailed Godwit during fat storage on a spring stopover site. Funct Ecol 17:151–159

Lee S-J (2004) Regulation of muscle mass by myostatin. Annu Rev Cell Dev Biol 20:61–86

Lee S-J, McPherron AC (2001) Regulation of myostatin activity and muscle growth. Proc Natl Acad Sci USA 98:9306–9311

Liknes ET (2005) Seasonal acclimatization patterns and mechanisms in small, temperate-resident passerines: phenotypic flexibility of complex traits. Ph.D. dissertation, University of South Dakota, Vermillion

Liknes ET, Swanson DL (1996) Seasonal variation in cold tolerance, basal metabolic rate, and maximal capacity for thermogenesis in White-breasted Nuthatches *Sitta carolinensis* and Downy Woodpeckers *Picoides pubescens*, two unrelated arboreal temperate residents. J Avian Biol 27:279–288

Liknes ET, Scott SM, Swanson DL (2002) Seasonal acclimatization in the American goldfinch revisited: to what extent to metabolic rates vary seasonally? Condor 104:548–557

Lindström Å (1997) Basal metabolic rates of migrating waders in the Eurasian Arctic. J Avian Biol 28:87–92

Lindström Å, Piersma T (1993) Mass changes in migrating birds: the evidence for fat and protein storage re-examined. Ibis 135:70–78

Lindström Å, Klaassen M, Kvist A (1999) Variation in energy intake and basal metabolic rate of a bird migrating in a wind tunnel. Funct Ecol 13:352–359

Lindström Å, Kvist A, Piersma T, Dekinga A, Dietz M (2000) Avian pectoral muscle size rapidly tracks body mass changes during flight, fasting and fuelling. J Exp Biol 203:913–919

Liu J-S, Li M (2006) Phenotypic flexibility of metabolic rate and organ masses among tree sparrows *Passer montanus* in seasonal acclimatization, Acta Zool. Sinica 52:469–477

López-Calleja MV, Bozinovic F (1995) Maximum metabolic rate, thermal insulation and aerobic scope in a small-sized Chilean hummingbird (*Sephanoides sephanoides*). Auk 112:1034–1036

Luiken JJFP, Schaap FG, van Nieuwenhoven FA, van der Vusse GJ, Bonen A, Glatz JFC (1999) Cellular fatty acid transport in heart and skeletal muscle as facilitated by proteins. Lipids 34:S169–S175

Lundgren BO (1988) Catabolic enzyme activities in the pectoralis muscle of migratory and non-migratory Goldcrests, Great Tits, and Yellowhammers. Ornis Scand 19:190–194

Lundgren BO, Kiessling K-H (1985) Seasonal variation in catabolic enzyme activities in breast muscle of some migratory birds. Oecologia 66:468–471

Lundgren BO, Kiessling K-H (1986) Catabolic enzyme activities in the pectoralis muscle of premigratory and migratory juvenile Reed Warblers *Acrocephalus scirpaceus* (Herm.). Oecologia 68:529–532

Lundgren BO, Kiessling K-H (1988) Comparative aspects of fibre types, areas, and capillary supply in the pectoralis muscle of some passerine birds with differing migratory behavior. J Comp Physiol B 158:165–173

Marjoniemi K, Hohtola E (1999) Shivering thermogenesis in leg and breast muscles of Galliform chicks and nestlings of the domestic pigeon. Physiol Biochem Zool 72:484–492

Marsh RL (1981) Catabolic enzyme activities in relation to premigratory fattening and muscle hypertrophy in the Gray Catbird (*Dumetella carolinensis*). J Comp Physiol 141:417–423

Marsh RL (1984) Adaptations of the Gray Catbird *Dumetella carolinensis* to long-distance migration: flight muscle hypertrophy associated with elevated body mass. Physiol Zool 57:105–117

Marsh RL (1993) Does regulatory nonshivering thermogenesis exist in birds? In: Carey C, Florant GL, Wunder BA, Horwitz B (eds) Life in the cold: ecological, physiological, and molecular mechanisms. Westview Press, Boulder, Colorado, pp 535–538

Marsh RL, Dawson WR (1982) Substrate metabolism in seasonally acclimatized American Goldfinches. Am J Physiol 242:R563–R569

Marsh RL, Dawson WR (1989) Avian adjustments to cold. In: Wang LCH (ed) Advances in comparative and environmental physiology 4: animal adaptation to cold. Springer, Berlin, pp 206–253

Marsh RL, Carey C, Dawson WR (1984) Substrate concentrations and turnover of plasma glucose during cold exposure in seasonally acclimatized house finches, *Carpodacus mexicanus*. J Comp Physiol B 154:469–476

Masman D, Gordijn M, Daan S, Dijkstra C (1986) Ecological energetics of the kestrel: field estimates of energy intake throughout the year. Ardea 74:24–39

Masman D, Daan S, Dietz M (1989) Heat increment of feeding in the kestrel, *Falco tinnunculus*, and its natural seasonal variation. In: Bech C, Reinertsen RE (eds) Physiology of cold adaptation in birds. Plenum Life Sciences, New York, pp 123–135

Mathieu-Costello O, Suarez RK, Hochachka PW (1992) Capillary-to-fiber geometry and mitochondrial density in hummingbird flight muscle. Respir Physiol 89:113–132

Mathieu-Costello O, Agey PJ, Logemann RB, Florez-Duquet M, Bernstein MH (1994) Effect of flying activity on capillary-fiber geometry in pigeon flight muscle. Tissue Cell 26:57–73

Mathieu-Costello O, Agey PJ, Quintana ES, Rousey K, Wu L, Bernstein MH (1998) Fiber capillarization and ultrastructure of pigeon pectoralis muscle after cold acclimation. J Exp Biol 201:3211–3220

Mawhinney K, Diamond AW, Kehoe FP (1999) The use of energy, fat, and protein reserves by breeding Great Black-backed Gulls. Can J Zool 77:1459–1464

McDonald RB, Horwitz BA, Stern JS (1988) Cold-induced thermogenesis in younger and older Fischer 344 rats following exercise training. Am J Physiol 254:R908–R916

McFarlan JT (2007) Seasonal upregulation of fatty acid uptake proteins in flight muscles of *Zonotrichia* sparrows during migration. M.S. Thesis, University of Western Ontario, London, ON

McKechnie AE (2008) Phenotypic flexibility in basal metabolic rate and the changing view of avian physiological diversity: a review. J Comp Physiol B 178:235–247

McKechnie AE, Lovegrove BG (2002) Avian facultative hypothermic responses: a review. Condor 104:705–724

McKechnie AE, Lovegrove BG (2006) Evolutionary and ecological determinants of avian torpor: a conceptual model. Acta Zoologica Sinica 52 (suppl):409–413

McKechnie AE, Wolf BO (2004) The allometry of avian basal metabolic rate: good predictions need good data. Physiol Biochem Zool 77:502–521

McKechnie AE, Freckleton RP, Jetz W (2006) Phenotypic plasticity in the scaling of avian metabolic rate. Proc R Soc B 1589:931–937

McNab BK (1988) Food habits and the basal rate of metabolism in birds. Oecologia 77:343–349

McNab BK (1997) On the utility of uniformity in the definition of basal rate of metabolism. Physiol Zool 70:718–720

McPherron AC, Lee S-J (1997) Double muscling in cattle due to mutations in the myostatin gene. Proc Natl Acad Sci USA 94:12457–12461

McWilliams SR, Karasov WH (2001) Phenotypic flexibility in digestive system structure and function in migratory birds and its ecological significance. Comp Biochem Physiol A 128:579–593

McWilliams SR, Caviedes-Vidal E, Karasov WH (1999) Digestive adjustments in Cedar Waxwings to high feeding rate. J Exp Zool 283:394–407

McWilliams SR, Guglielmo C, Pierce B, Klaassen M (2004) Flying, fasting, and feeding in birds during migration: a nutritional and physiological ecology perspective. J Avian Biol 35:377–393

Merola-Zwartjes M (1998) Metabolic rate, temperature regulation, and the energetic implications of roost nests in the Bananaquit (*Coereba flaveola*). Auk 115:780–786

Mezentseva NV, Kumaratilake JS, Newman SA (2008) The brown adipocyte differentiation pathway in birds: an evolutionary road not taken. BMC Biol 6:17. doi:10.1186/1741-7007-6-17

Miller DS (1939) A study of the physiology of the sparrow thyroid. J Exp Zool 80:259–281

Moe B, Langseth I, Fyhn M, Gabrielsen GW, Bech C (2002) Changes in body condition in breeding kittiwakes *Rissa tridactyla*. J Avian Biol 33:225–234

Moreno J (1989) Strategies of mass change in breeding birds. Biol J Linn Soc 37:297–310

Morris SR, Richmond ME, Holmes DW (1994) Patterns of stopover by warblers during spring and fall migration on Appledore Island, Maine. Wilson Bull 106:703–718

Mozo J, Emre Y, Bouillaud F, Ricquier D, Criscuolo F (2005) Thermoregulation: what role for UCPs in mammals and birds? Biosci Rep 25:227–249

Murphy MT, Armbrecth B, Vlamis E, Pierce A (2000) Is reproduction by tree swallows cost free? Auk 117:902–912

Nagy LR, Stanculescu D, Holmes RT (2007) Mass loss by breeding female songbirds: food supplementation supports energetic stress hypothesis in Black-throated Blue Warblers. Condor 109:304–311

Newton I, Dale LC (1996) Bird migration at different latitudes in eastern North America. Auk 113:626–635

Nilsson J-A, Råberg L (2001) The resting metabolic cost of egg laying and nestling feeding in great tits. Oecologia 128:187–192

Norberg Å (1981) Temporary weight decrease in breeding birds may result in more fledged young. Am Nat 118:838–850

Novoa FF, Veloso C, López-Calleja MV, Bozinovic F (1996) Seasonal changes in diet, digestive morphology and digestive efficiency in the Rufous-collared Sparrow (*Zonotrichia capensis*) in central Chile. Condor 98:873–876

Nudds RL, Bryant DM (2000) The energetic cost of short flights in birds. J Exp Biol 203:1561–1572

Nur N (1984) The consequences of brood size for breeding Blue Tits I. Adult survival, weight change and the cost of reproduction. J Anim Ecol 53:479–496

O'Connor TP (1995a) Metabolic characteristics and body composition in house finches: effects of seasonal acclimatization. J Comp Physiol B 165:298–305

O'Connor TP (1995b) Seasonal acclimatization of lipid mobilization and catabolism in House Finches (*Carpodacus mexicanus*). Physiol Zool 68:985–1005

O'Reilly KM, Wingfield JM (1995) Spring and autumn migration in Arctic shorebirds: same distance, different strategies. Am Zool 35:222–233

Olson JM, Dawson WR, Camilliere JJ (1988) Fat from Black-capped Chickadees: avian brown adipose tissue? Condor 90:529–537

Palacios L, Palomeque J, Riera M, Pages T, Viscor G, Planas J (1984) Oxygen transport properties in the starling, *Sturnus vulgaris* L. Comp Biochem Physiol 77A:255–260

Paladino FV, King JR (1984) Thermoregulation and oxygen consumption during terrestrial locomotion by White-crowned Sparrows, *Zonotricia leucophrys gambelii*. Physiol Zool 57:226–236

Pearson DJ, Lack PC (1992) Migration patterns and habitat use by passerine and near- passerine migrant birds in eastern Africa. Ibis 134 (suppl):89–98

Pelsers MMAL, Butler PJ, Bishop CM, Glatz JFC (1999) Fatty acid binding protein in heart and skeletal muscles of the migratory Barnacle Goose throughout development. Am J Physiol 276:R637–R643

Phillips RA, Furness RW (1997) Sex-specific loss of mass by breeding Arctic Skuas. J Avian Biol 28:163–170

Pierce BJ, McWilliams SR, O'Connor TP, Place AR, Guglielmo CG (2005) Effect of dietary fatty acid composition on depot fat and exercise performance in a migrating songbird, the red-eyed vireo. J Exp Biol 208:1277–1285

Piersma T (1998) Phenotypic flexibility during migration: optimization of organ size contingent on the risks and rewards of fueling and flight? J Avian Biol 29:511–520

Piersma T (2002) Energetic bottlenecks and other design constraints in avian annual cycles. Integ Comp Biol 42:51–67

Piersma T, Dietz MW (2007) Twofold seasonal variation in the supposedly constant, species-specific, ratio of upstroke to downstroke flight muscles in Red Knots *Calidris canutus*. J Avian Biol 35:536–540

Piersma T, Gill RE Jr (1998) Guts don't fly: small digestive organs in obese Bar-tailed Godwits. Auk 115:196–203

Piersma T, Jukema J (2002) Contrast in adaptive mass gains: Eurasian Golden Plovers store fat before midwinter and protein before prebreeding flight. Proc R Soc Lond B 269:1101–1105

Piersma T, Cadée N, Daan S (1995) Seasonality in basal metabolic rate and thermal conductance in a long-distance migrant shorebird, the knot (*Calidris canutus*). J Comp Physiol B 165:37–45

Piersma T, Bruinzeel L, Drent R, Kersten M, Van der Meer J, Wiersma P (1996a) Variability in basal metabolic rate of a long-distance migrant shorebird (Knot *Calidris canutus*) reflects shifts in organ size. Physiol Zool 69:191–217

Piersma T, Everaarts JM, Jukema J (1996b) Build-up of red blood cells in refueling Bar-tailed Godwits in relation to individual migratory quality. Condor 98:363–370

Piersma T, Gudmundsson GA, Lilliendahl K (1999) Rapid changes in the size of different functional organ and muscle groups during refueling in a long-distance migrating shorebird. Physiol Biochem Zool 72:405–415

Piersma T, Koolhaas A, Dekinga A, Gwinner E (2000) Red blood cell and white blood cell counts in sandpipers (*Philomachus pugnax, Calidris canutus*): effects of captivity, season, nutritional status, and frequent bleedings. Can J Zool 78:1349–1355

Piersma T, Gessaman JA, Dekinga A, Visser GH (2004) Gizzard and other lean mass components increase, yet basal metabolic rates decrease, when Red Knots *Calidris canutus* are shifted from soft to hard-shelled food. J Avian Biol 35:99–104

Raimbault S, Dridi S, Denjean F, Lachuer J, Couplan E, Bouillaud F, Bordas A, Duchamp C, Taouis M, Ricquier D (2001) An uncoupling protein homologue putatively involved in facultative muscle thermogenesis in birds. Biochem J 353:441–444

Ramenofsky M (1990) Fat storage and fat metabolism in relation to migration. In: Gwinner E (ed) Bird migration: physiology and ecophysiology. Springer, Berlin, pp 214–231

Ramenofsky M, Savard R, Greenwood MRC (1999) Seasonal and diel transitions in physiology and behavior in the migratory dark-eyed junco. Comp Biochem Physiol A 122:385–397

Rashotte ME, Saarela S, Henderson RP, Hohtola E (1999) Shivering and digestion-related thermogenesis in pigeons during dark phase. Am J Physiol 277:R1579–R1587

Reinertsen RE (1996) Physiological and ecological aspects of hibernation. In: Carey C (ed) Avian energetics and nutritional ecology. Chapman & Hall, New York, pp 125–157

Rezende EL, Swanson DL, Novoa FF, Bozinovic F (2002) Passerines versus nonpasserines: so far, no statistical differences in the scaling of avian energetics. J Exp Biol 205:101–107

Ricklefs RE, Hussell DJT (1984) Changes in adult mass associated with the nesting cycle in the European Starling. Ornis Scand 15:155–161

Roberts TJ, Weber J-M, Hoppeler H, Weibel EW, Taylor CR (1996) Design of the oxygen and substrate pathways, II. Defining the upper limits of carbohydrate and fat oxidation. J Exp Biol 199:1651–1658

Rønning B, Moe B, Chastel O, Broggi J, Langset M, Bech C (2008) Metabolic adjustments in breeding female kittiwakes (*Rissa tridactyla*) include changes in kidney metabolic intensity. J Comp Physiol B 178:779–784

Rosenmann M, Morrison P (1974) Maximum oxygen consumption and heat loss facilitation in small homeotherms by He-O$_2$. Am J Physiol 226:490–495

Saarela S, Hissa R, Pyörnilä A, Harjula R, Ojanen M, Orell M (1989a) Do birds possess brown adipose tissue? Comp Biochem Physiol 92A:219–228

Saarela S, Klapper B, Heldmaier G (1989b) Thermogenic capacity of greenfinches and siskins in winter and summer. In: Bech C, Reinertsen RE (eds) Physiology of cold adaptation in birds. Plenum, New York, pp 115–122

Saarela S, Keith JS, Hohtola E, Trayhurn P (1991) Is the "mammalian" brown fat-specific mitochondrial uncoupling protein present in adipose tissues of birds? Comp Biochem Physiol 100B:45–49

Saarela S, Klapper B, Heldmaier G (1995) Daily rhythm of oxygen consumption and thermoregulatory responses in some European winter- or summer-acclimatized finches at different ambient temperatures. J Comp Physiol B 165:366–376

Sanz JJ, Moreno J (1995) Mass loss in brooding female Pied Flycatchers *Ficedula hypoleuca*: no evidence for reproductive stress. J Avian Biol 26:313–320

Saunders DK, Fedde MR (1991) Physical conditioning: effect on the myoglobin concentration in skeletal and cardiac muscle of Bar-headed Geese. Comp Biochem Physiol 100A:349–352

Savard R, Ramenofsky M, Greenwood MRC (1991) A north-temperate migratory bird: a model for the fate of lipids during exercise of long duration. Can J Physiol Pharmacol 69:1443–1447

Schwilch R, Jenni L, Jenni-Eiermann S (1996) Metabolic responses of homing pigeons to flight and subsequent recovery. J Comp Physiol B 166:77–87

Scott I, Mitchell PI, Evans PR (1996) How does variation in body composition affect the basal metabolic rate of birds? Funct Ecol 10:307–313

Seymour RS, Runciman S, Baudinette RV (2008) Development of maximum metabolic rate and pulmonary diffusing capacity in the superprecocial Australian Brush Turkey *Alectura lathami*: an allometric and morphometric study. Comp Biochem Physiol A 150:169–175

Speakman JR (1997) Doubly labelled water theory and practice. Chapman & Hall, London

Stevens L (1996) Avian biochemistry and molecular biology. Cambridge University Press, Cambridge

Strømme SB, Hammel HT (1967) Effects of physical training on tolerance to cold in rats. J Appl Physiol 23:815–824

Suarez RK (1998) Oxygen and the upper limits to animal design and performance. J Exp Biol 201:1065–1072

Swanson DL (1990a) Seasonal variation in cold hardiness and peak rates of cold-induced thermogenesis in the Dark-eyed Junco (*Junco hyemalis*). Auk 107:561–566

Swanson DL (1990b) Seasonal variation of vascular oxygen transport in the Dark-eyed Junco. Condor 92:62–66

Swanson DL (1991a) Seasonal adjustments in metabolism and insulation in the Dark-eyed Junco. Condor 93:538–545

Swanson DL (1991b) Substrate metabolism under cold stress in seasonally acclimatized Dark-eyed Juncos. Physiol Zool 64:1578–1592

Swanson DL (1993) Cold tolerance and thermogenic capacity in Dark-eyed Juncos in winter: geographic variation and comparison with American Tree Sparrows. J Therm Biol 18:275–281

Swanson DL (1995) Seasonal variation in thermogenic capacity of migratory Warbling Vireos. Auk 112:870–877

Swanson DL (2001) Are summit metabolism and thermogenic endurance correlated in winter acclimatized passerine birds? J Comp Physiol B 171:475–481

Swanson DL, Dean KL (1999) Migration-induced variation in thermogenic capacity in migratory passerines. J Avian Biol 30:245–254

Swanson DL, Liknes ET (2006) A comparative analysis of thermogenic capacity and cold tolerance in small birds. J Exp Biol 209:466–474

Swanson DL, Olmstead KL (1999) Evidence for a proximate influence of winter temperature on metabolism in passerine birds. Physiol Biochem Zool 72:566–575

Swanson DL, Weinacht DP (1997) Seasonal effects on metabolism and thermoregulation in Northern Bobwhite. Condor 99:478–489

Swanson DL, Sabirzhanov B, VandeZande A, Clark TG (2009) Down-regulation of myostatin gene expression in pectoralis muscle is associated with elevated thermogenic capacity in winter acclimatized house sparrows (*Passer domesticus*). Physiol Biochem Zool 82:121–128

Sweazea KL, Braun EJ (2006) Oleic acid uptake by in vitro English Sparrow muscle. J Exp Zool 305A:268–276

Taigen TL (1983) Activity metabolism of anuran amphibians: implication for the origin of endothermy. Am Nat 121:94–109

Tatner P, Bryant DM (1986) Flight cost of a small passerine measured using doubly labeled water: implications for energetics studies. Auk 103:169–180

Taylor CR, Weibel ER (1981) Design of the mammalian respiratory system. I. Problem and strategy. Respir Physiol 44:1–10

Tøien O (1992) Data acquisition in thermal physiology: measurements of shivering. J Therm Biol 17:357–366

Trevelyan R, Harvey PH, Pagel MD (1990) Metabolic rates and life histories in birds. Funct Ecol 4:135–141

Turner DL, Hoppeler H, Hokanson J, Weibel ER (1995) Cold acclimation and endurance training in guinea pigs: changes in daily and maximal metabolism. Respir Physiol 101:183–188

Vaillancourt E, Prud'Homme S, Haman F, Guglielmo CG, Weber J-M (2005) Energetics of a long-distance migrant shorebird (*Philomachus pugnax*) during cold exposure and running. J Exp Biol 208:317–325

Vezina F, Williams TD (2002) Metabolic cost of egg production in the European Starling (*Sturnus vulgaris*). Physiol Biochem Zool 75:377–385

Vezina F, Williams TD (2003) Plasticity in body composition in breeding birds: what drives the metabolic costs of egg production? Physiol Biochem Zool 76:716–730

Vezina F, Williams TD (2005) Interaction between organ mass and citrate synthase activity as an indicator of tissue maximal oxidative capacity in breeding European Starlings: implications for metabolic rate and organ mass relationships. Funct Ecol 19:119–128

Vezina F, Jalvingh KM, Dekinga A, Piersma T (2006) Acclimation to different thermal conditions in a northerly wintering shorebird is driven by body mass-related changes in organ size. J Exp Biol 209:3141–3154

Vezina F, Jalvingh KM, Dekinga A, Piersma T (2007) Thermogenic side effects to migratory disposition in shorebirds. Am J Physiol Regul Integr Comp Physiol 292:1287–1297

Vianna CR, Hagen T, Zhang C-Y, Bachman E, Boss O, Gereben B, Moriscot AS, Lowell BB, Bicudo JEPW, Bianco AC (2001) Cloning and functional characterization of an uncoupling protein homolog in hummingbirds. Physiol. Genomics 5:137–145

Vittoria JC, Marsh RL (1996) Cold-acclimated ducklings shiver when exposed to cold. Am Zool 36:66A

Vleck CM, Vleck D (1979) Metabolic rate in five tropical bird species. Condor 81:89–91

Weathers WW (1979) Climatic adaptation in avian standard metabolic rate. Oecologia 42:81–89

Weathers WW (1997) Energetics and thermoregulation by small passerines of the humid lowland tropics. Auk 114:341–353

Weathers WW, Caccamise DR (1978) Seasonal acclimatization to temperature in Monk Parakeets. Oecologia 35:173–183

Weathers WW, Sullivan KA (1993) Seasonal patterns of time and energy allocation by birds. Physiol Zool 66:511–536

Weathers WW, Olson CR, Siegel RB, Davidson CL, Famula TR (1999) Winter and breeding-season energetics of nonmigratory White-crowned Sparrows. Auk 116:842–847

Weber TP, Piersma T (1996) Basal metabolic rate and the mass of tissues differing in metabolic scope: migration-related covariation between individual knots, *Calidris canutus*. J Avian Biol 27:215–224

Weber J-M, Brichon G, Zwingelstein G, McClelland G, Saucedo C, Weibel ER, Taylor CR (1996) Design of the oxygen and substrate pathways, IV. Partitioning energy provision from fatty acids. J Exp Biol 199:1667–1674

Webster MD, Weathers WW (1990) Heat produced as a by-product of foraging activity contributes to thermoregulation by verdins, *Auriparus flaviceps*. Physiol Zool 63:777–794

Webster MD, Weathers WW (2000) Seasonal changes in energy and water use by Verdins, *Auriparus flaviceps*. J Exp Biol 203:3333–3344

White CR, Blackburn TM, Martin GR, Butler PJ (2007) Basal metabolic rate of birds is associated with habitat temperature and precipitation, not primary productivity. Proc R Soc B 274:287–293

Wiersma P, Piersma T (1994) Effects of microhabitat, flocking, climate and migratory goal on energy expenditure in the annual cycle of Red Knots. Condor 96:257–279

Wiersma P, Muñoz-Garcia A, Walker A, Williams JB (2007a) Tropical birds have a slow pace of life. Proc Natl Acad Sci USA 104:9340–9345

Wiersma P, Chappell MA, Williams JB (2007b) Cold-and exercise-induced peak metabolic rates in tropical birds. Proc Natl Acad Sci USA 104:20866–20871

Williams JB (2001) Energy expenditure and water flux of free-living Dune Larks in the Namib: a test of the reallocation hypothesis on a desert bird. Funct Ecol 15:175–185

Williams JB, Tieleman BI (2000) Flexibility in basal metabolic rate and evaporative water loss among Hoopoe Larks exposed to different environmental temperatures. J Exp Biol 203:3153–3159

Williams CT, Kildaw SD, Buck CL (2007) Sex-specific differences in body condition indices and seasonal mass loss in Tufted Puffins. J Field Ornithol 78:369–378

Winker K, Warner DW, Weisbrod AR (1992) Timing of songbird migration in the St. Croix River Valley, Minnesota, 1984-1986. Loon 64:131–137

Wolfman NM, McPherron AC, Pappano WN, Davies MV, Song K, Tomlinson KN, Wright JF, Zhao L, Sebald SM, Greenspan DS, Lee S-J (2003) Activation of latent myostatin by the BMP-1/tolloid family of metalloproteinases. Proc Natl Acad Sci USA 100:15842–15846

Yacoe ME, Dawson WR (1983) Seasonal acclimatization in American goldfinches: the role of the pectoralis muscle. Am J Physiol 245:R265–R271

Zerba E, Walsberg GE (1992) Exercise-generated heat contributes to thermoregulation by Gambel's Quail in the cold. J Exp Biol 171:409–422

Zerba E, Dana AN, Lucia MA (1999) The influence of wind and locomotor activity on surface temperature and energy expenditure of the eastern House Finch (*Carpodacus mexicanus*) during cold stress. Physiol Biochem Zool 72:265–276

Zheng W-H, Li M, Liu J-S, Shao S-L (2008) Seasonal acclimatization of metabolism in Eurasian Tree Sparrows (*Passer montanus*). Comp Biochem Physiol A 151:519–525

Chapter 4
Assessing Cause–Effect Relationships in Environmental Accidents: Harlequin Ducks and the *Exxon Valdez* Oil Spill

John A. Wiens, Robert H. Day, Stephen M. Murphy, and Mark A. Fraker

4.1 Introduction

Over the past several decades, scientists, resource managers, politicians, policy makers, and the general public have all become increasingly aware of the magnitude and complexity of environmental problems resulting from human activities. Assessing the ecological consequences of environmental disruptions such as habitat fragmentation, timber harvesting, or the release of pollutants into the environment has become the focus of entire disciplines. In all of these fields, the development of remediation, mitigation, or management strategies ultimately rests on assessing cause–effect relationships.

Consider oil spills as an example. In marine ecosystems, oil spills can have severe impacts on wildlife, especially for species that are closely tied to the near-shore environment, where oil often accumulates after a spill. Thus, when differences in abundance, survival, reproduction, habitat use, or other attributes of individuals or populations are observed between oiled and unoiled areas after an oil spill, it is easy to conclude that these differences are caused by the spill. The logic behind this conclusion is clear, and it accords well with our preconception that oil spills are bad for wildlife.

Unfortunately, simple logic does not always match reality. Assessing cause–effect linkages following an oil spill is not as straightforward as it may appear. On the one hand, preconceptions, emotions, unproven assumptions, and advocacy may stand in the way of reaching reliable conclusions. On the other hand, constraints on study design or statistical analysis may compromise the rigor and increase the uncertainty of conclusions based on scientific studies of such events. Like most environmental accidents, oil spills are not random events in either space or time, so the core assumption of many statistical methods is suspect (Wiens and Parker 1995). The broad issue that confronts any scientist, then, is how to determine with a

J.A. Wiens (✉)
PRBO Conservation Science, 3820 Cypress Drive #11, Petaluma, CA 94954, USA
e-mail: jwiens@prbo.org

C.F. Thompson (ed.), *Current Ornithology Volume 17*, Current Ornithology,
DOI 10.1007/978-1-4419-6421-2_4, © Springer Science+Business Media, LLC 2010

reasonable degree of rigor and certainty what happened following an environmental disturbance, whether it be an oil spill, habitat fragmentation, timber harvesting, or some other anthropogenic change. Reliably assessing these effects is critical, not only to scientists who are trying to understand patterns of ecological response to perturbations and to managers attempting to mitigate and restore damaged resources, but also to the litigation that inevitably follows such events.

To explore some of the difficulties in assessing causation and developing scientifically rigorous conclusions about the consequences of environmental disturbances, we focus here on a particularly instructive case study: Harlequin Ducks (*Histrionicus histrionicus*) and the *Exxon Valdez* oil spill. This focus is appropriate because (1) the *Exxon Valdez* oil spill is the largest tanker spill to occur in North American waters; (2) it caused the immediate mortality of hundreds of thousands of seabirds, including Harlequin Ducks (Piatt and Ford 1996; Piatt et al. 1990); (3) the Harlequin Duck is a charismatic species of conservation concern; (4) it is closely associated with shoreline environments, where it may be especially vulnerable to oiling (King and Sanger 1979); and (5) it has been identified as a species that has not yet fully recovered from the *Exxon Valdez* oil spill [*Exxon Valdez* Oil Spill Trustee Council (hereafter, Trustees), 2009; status last updated 2009, see http://www.evostc.state.ak.us/publications/annualstatus.cfm]. We adopt this specific focus because it allows us to examine the difficulties in assessing cause and effect. We make no attempt to review either the large body of literature that deals with the effects of oil spills or other environmental contaminants on birds (see Burger 1997; Frink et al. 1995; Wiens 1995; Furness and Greenwood 1993; Hunt 1987; Clark 1984) or the effects of the *Exxon Valdez* oil spill on other elements of the marine ecosystem (see National Wildlife Federation 2003; Carls et al. 2002; Peterson 2001; Rice et al. 1996; Wells et al. 1995; Loughlin 1994), both of which are beyond the scope of this evaluation. Esler et al. (2002) have also reviewed the effects of the *Exxon Valdez* oil spill on Harlequin Ducks but have done so from a somewhat different perspective.

In this case study, we begin by providing some essential background on features of the Prince William Sound (hereafter, PWS) environment, of Harlequin Duck ecology, and of the *Exxon Valdez* oil spill. We then review statements that have been made regarding the effects of this spill on Harlequins. We follow with a more detailed examination of the evidence: what do we really know, and what can we unequivocally say, about the effects of the spill on Harlequin Duck distribution and abundance, reproduction and survival, and habitat use? What alternative or complementary hypotheses could explain the observations? What approaches have been used to examine cause–effect relationships, what do they show, and how adequate are they? We conclude by considering the broader implications of this case study for the assessment of the impacts of environmental disruptions. How can reliable conclusions be reached when such disruptions are essentially unplanned and poorly designed "experiments?" Finally, how can uncertain knowledge be balanced with the need to determine impacts and recovery and to initiate mitigation or restoration efforts?

4.2 Background

4.2.1 The Prince William Sound Environment

PWS is a large embayment of the Northern Gulf of Alaska that includes numerous fjords, islands, islets, and reefs (Fig. 4.1). The coastline is highly complex, with ~4,800 km of shoreline in an area only ~125 km in diameter (Isleib and Kessel 1973). Most of the northern part of PWS is either glaciated or has recently been deglaciated, resulting in primarily rocky shorelines in that part of the Sound. In contrast, some of southern PWS (especially Montague Island) has numerous wide, finer-grained beaches as a result of the Alaska earthquake of 1964 (Isleib and Kessel 1973). Eastern PWS is less rugged and rocky and is more heavily forested than is western PWS. In western PWS, Knight Island and nearby areas are particularly rugged, with mostly rocky shorelines, few large streams, and numerous cliffs. Marine productivity is lower in the glacial waters of northern PWS than in the warmer waters of the Alaska Coastal Current, which flows through the southern Sound (Royer et al. 1990).

4.2.2 Harlequin Duck Ecology

Harlequin Ducks are small seaducks that breed along clear, fast-flowing rivers and streams and winter in shallow intertidal zones of rocky coastlines. They exhibit delayed sexual maturity, low annual recruitment, and high adult survivorship (Rosenberg and Petrula 1998; Goudie et al. 1994b) – all features of species adapted to harsh and/or highly variable environments. General aspects of their ecology, distribution, behavior, and life history have been reviewed by Robertson and Goudie (1999).

In coastal PWS, Harlequin Ducks occur throughout the year, although many individuals leave the coastal areas in the spring to breed in inland nesting areas and then return in late summer and fall (Rosenberg and Petrula 1998; Isleib and Kessel 1973). Densities peak in the fall, when both resident and migrant birds are present, then decline through the rest of the winter. In spring and summer, prebreeding birds concentrate near the mouths of salmon-spawning streams, which are preferred nesting habitat (Robertson and Goudie 1999; Crowley 1994; Dzinbal 1982). Birds move as far as several kilometers upstream to nest. Egg-laying occurs from mid-May to mid-June, and hatching occurs from mid-June onward. Initially, broods are reared on their natal streams and adjacent freshwater habitats, but they soon move to nearby estuaries and protected shorelines in late July to mid-August to complete growth (Robertson and Goudie 1999; Dzinbal 1982; Sangster et al. 1978; Nysewander and Knudtson 1977; Isleib and Kessel 1973; Wiens et al., unpublished data). Harlequin Duck broods become fully feathered at ~45 days of age and flight-capable at ~55 days, although the timing to independence varies both within

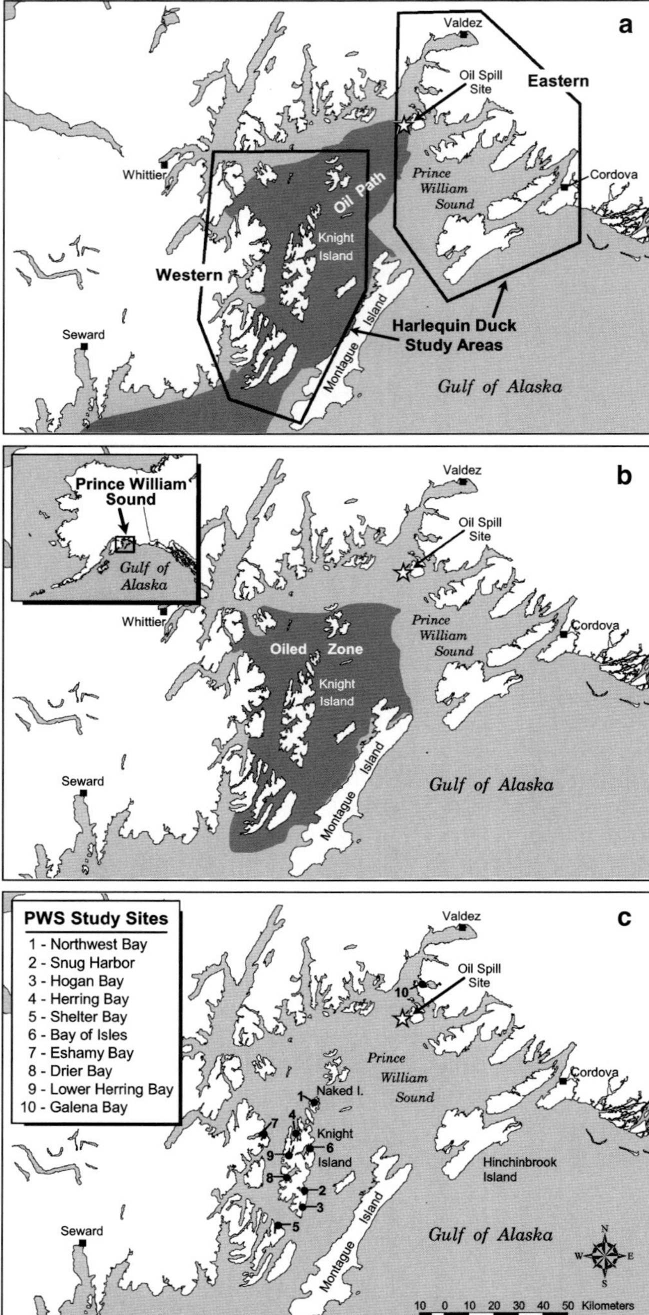

Fig. 4.1 Prince William Sound, Alaska, showing (**a**) the path of the *Exxon Valdez* oil spill (from Wiens 1996) and the designations of eastern and western regions used in several studies to compare unoiled with oiled areas (from Patten et al. 2000); (**b**) the designation of the oiled and unoiled regions of the Sound used by Klosiewski and Laing (1994); and (**c**) the ten bays used in a fine-scale analysis of spill effects (from Murphy et al. 1997)

and among populations (Robertson and Goudie 1999). From mid-July to mid-August, male and nonbreeding female Harlequins molt are flightless, remaining within a few meters of shorelines in sheltered bays; in contrast, breeding females arrest the molt until after the young have fledged. Adult Harlequin Ducks may exhibit strong site fidelity to both breeding (Cassirer and Groves 1994; Kuchel 1977; Bengtson 1972) and molting/wintering locations (Iverson and Esler 2006; Iverson et al. 2004; Cooke et al. 2000; Robertson et al. 2000), whereas site fidelity is less pronounced in subadult females and juveniles of both sexes (Holland-Bartels et al. 1998).

In the fall and winter, Harlequin Ducks typically loaf on small offshore rocks and forage around rocky shorelines, where they feed in shallow waters on a wide range of intertidal invertebrates, primarily gastropods (especially *Littorina* snails) and crustaceans such as hermit crabs and amphipods (Fischer and Griffin 2000; Patten et al. 2000; see also Robertson and Goudie 1999; Vermeer 1983). In the winter and early spring, individuals may aggregate in considerable numbers around localized, ephemeral food sources such as herring spawn, with some individuals moving as much as 80 km to join such groups (Rodway et al. 2003a). In the spring and early summer, Harlequins feed on a variety of invertebrates (*Littorina* and other snails, shore and hermit crabs, sea stars, insects) and eelgrass (*Zostera*) in intertidal areas of protected marine waters and at the mouths of small streams, shifting to forage on drifting salmon roe in lower portions of salmon-spawning streams later in the summer (Dzinbal and Jarvis 1984). Young forage primarily on dipteran larvae in eddies and slower moving portions of streams; in some locations, reproductive success is markedly reduced in years or areas in which insect prey are scarce in streams (Bengtson 1972; Bengtson and Ulfstrand 1971).

By nature of their life-history traits and ecology, Harlequin Ducks are potentially among the most vulnerable species to oil spills (Esler et al. 2000a; Holland-Bartels et al. 1998; King and Sanger 1979). Day et al. (1997, 1995) found that the magnitude of impacts of the *Exxon Valdez* oil spill on bird species occurring in marine and shoreline habitats in PWS was greatest for species that foraged in intertidal and nearshore habitats and for species that were year-around residents. The tendency of Harlequins to forage in intertidal areas, their residency in PWS, and their tendency to loaf on intertidal rocks and spend time near shorelines all enhance their potential vulnerability to an oil spill.

4.2.3 *The Exxon Valdez Oil Spill*

The details of the *Exxon Valdez* oil spill have been well documented (Rice et al. 1996; Wells et al. 1995; Wheelwright 1994). The tanker grounded on Bligh Reef on 24 Mar 1989, releasing ~41,000,000 L of Alaska North Slope crude oil into the waters of northeastern PWS. The oil drifted to the west and southwest from the spill site, generally affecting shorelines in the western, but not the eastern, portion of PWS ("Western" and "Eastern" in Fig. 4.1a). Within western PWS, however, the level of oiling of shorelines varied considerably, even within short distances.

About 783 km (~16%) of the ~4,800 km of shoreline in PWS were oiled (Boehm et al. 2008; Neff et al. 1995). Shores on the eastern and northern sides of the Knight Island Group and nearby islands generally received much oil, whereas shores on the western and southern sides were more protected and received little or no oil. The concentration of polycyclic aromatic hydrocarbons (PAH's) from *Exxon Valdez* oil in the water column had returned to background levels at most locations by June 1989 and at all locations by the spring of 1990 (Neff and Stubblefield 1995; Wolfe et al. 1994), and the amount of oil on and in beach sediments declined dramatically between 1989 and 1992 (Neff et al. 1995; Wolfe et al. 1994) and continued to decline afterward (Boehm et al. 2008; Page et al. 2008, 2005, 1999, Short et al. 2002, 2004, 2006, 2007; Hoff and Shigenaka 1999). The toxicity of oil in beach sediments also decreased rapidly as the lighter and more toxic fractions dissipated and fell below toxic thresholds in all but a few locations by 1990–1991 (Page et al. 2002a, 2002b; Boehm et al. 1995; but see Boehm et al. 2008; Page et al. 2003; Rice et al. 2003). Cleanup operations were extensive in 1989, considerably reduced in 1990, and minor in 1991, as the quantity of beached oil declined as a result of the cleanup operations and natural weathering. By 1993, the extent of oiled shoreline had decreased from ~800 to ~14 km, 92% of which was very lightly oiled (i.e., the intertidal zone had ≤10% oil cover); heavily oiled shoreline (i.e., the intertidal zone had a band of oil >6 m wide and ≥50% oil cover) had declined from 140 km in 1989 to 0.1 km in 1993 (Neff et al. 1995). A separate, Trustee-sponsored survey of all of the originally oiled areas in PWS found subsurface oil residues remaining in widely scattered locations along about 7 km of shoreline and surficial oil residues along 4.8 km of shoreline in 1993 (Peterson 2001; Gibeaut and Piper 1998). Although the quantity and distribution of residual oil continue to be debated, it is clear that some oil residues persisted in a moderately unweathered and potentially toxic state in isolated locations in PWS several years after the spill (Hayes and Michel 1999). Based on a stratified random survey conducted during 2001 (primarily on sites that had been heavily oiled in 1989 and that had substantial oil residues remaining in 1990–1993), Short et al. (2004, 2002) estimated that approximately 11.3 ha of shoreline (equivalent to ~5.8 km of linear shoreline) in PWS were still contaminated with surface and subsurface oil residues. In this and more recent surveys, most remaining oil residues were highly weathered (81% of subsurface oil samples analyzed from a 2007 survey had >70% loss of total PAH with respect to the cargo crude; Boehm et al. 2008) and confined to widely scattered patches in the middle and upper intertidal zones, sequestered under a surface armor of boulders and cobbles (Boehm et al. 2008; Taylor and Reimer 2008).

4.3 Concerns About Impacts to Harlequin Ducks

Within weeks of the spill, concerns were expressed about detrimental effects on numerous marine bird species, including Harlequin Ducks. To provide a context for the examination of evidence that follows, we give a sample of the statements made and conclusions drawn about spill impacts on Harlequin Ducks in Table 4.1. These

Table 4.1 A sampling of statements made and conclusions drawn in the peer-reviewed scientific literature, in reports, and in the public media about the effects of the *Exxon Valdez* oil spill on Harlequin Ducks in Prince William Sound, arranged chronologically

Harlequin Ducks "may have suffered high losses relative to the size of the local populations" (Piatt et al. 1990)

Harlequin Ducks in the unoiled portion of Prince William Sound were in "much better condition" than birds in the spill area (Patten 1993a)

"Since the spill, ducks living in areas hit by oil have failed to breed at all" (Pain 1993)

It is "possible that a local extinction of Harlequin Ducks may occur within the spill area" (Patten 1993a)

"We find it unlikely that these (population) declines were caused only by the oil spill" (Klosiewski and Laing 1994)

"Overall, definitive conclusions cannot be drawn about the condition of harlequin ducks in western Prince William Sound" (Patten 1994)

"Densities of Harlequins were consistently low (in 1990–1992) in oiled areas" (Patten 1994)

"Breeding population of Prince William Sound, AK, decimated by *Exxon Valdez* oil spill" (Goudie et al. 1994a)

"Direct mortality of females, combined with sublethal effects of oil toxicity on reproductive physiology and survival likely caused low productivity and decline of molting harlequin ducks" (Patten et al. 1995)

"Considering the low frequency of mussel beds with residual oil, the patchy distribution of remaining weathered oil residues, and the relatively low PAH concentrations in the mussels, the risk of quantifiable injury at the level of an individual bird ... or at the population level, is minimal" (Boehm et al. 1996)

"Harlequin Duck (habitat use) was negatively impacted, with evidence of recovery seen by 1991" (Day et al. 1997)

"A negative trend in harlequin abundance in western (oiled) Prince William Sound and a positive trend in eastern (unoiled) Prince William Sound suggests that harlequin numbers in western Prince William Sound are still declining" (Rosenberg and Petrula 1998)

"There continues to be concern about poor reproduction and survival in oiled areas, although the overall population in Prince William Sound appears to be increasing" (Exxon Valdez Oil Spill Trustee Council 1998)

The "breeding propensity of female harlequin ducks is lower in western than in eastern Prince William Sound. However, the presence of a relatively large number of sub-adults in western Prince William Sound suggests that the lack of breeding activity by females in that region may be the result of females not attaining breeding age, and not because they have failed to attain breeding condition" (Rosenberg and Petrula 1998)

"We suspect relatively little breeding habitat exists in Prince William Sound, but relatively more is available in eastern Prince William Sound" (Rosenberg and Petrula 1998)

"Harlequin ducks exhibited negative population effects in 1990 and 1991 only" (Irons et al. 1999)

"It is clear that the breeding habitat in the western sound is very limited compared to what is available in the eastern sound ... conclusions of reproductive failure based on lack of broods in the oiled area do not now seem warranted" (Exxon Valdez Oil Spill Trustee Council 1999)

"Many intertidal mussel beds are still contaminated with oil, seriously handicapping recovery of sea otters and harlequin ducks" (Ott 1999)

"Lack of recovery of Harlequin Ducks in oiled areas does not appear to be related to their population-genetic characteristics, but may instead be associated with unfavorable local environmental conditions" (Lanctot et al. 1999)

"The effects are still there on various populations. The populations are diminished in the spill area. They have not recovered. But you have to know where to look and what you're looking for" (McCammon 1999)

(continued)

Table 4.1 (continued)

"… Full recovery of some sea duck populations impacted by the *Exxon-Valdez* oil spill may be constrained by exposure to residual oil" (Trust et al. 2000)

"Survival of (adult female harlequin ducks) in oiled areas was lower than in unoiled areas … continued effects of the oil spill likely restricted recovery of harlequin duck populations through at least 1998" (Esler et al. 2000a)

After accounting for habitat effects, "densities were lower in oiled than unoiled areas, suggesting that population recovery from the oil spill was not complete" (Esler et al. 2000b)

"Harlequin Ducks … displayed strong evidence of negative oil spill effects a few years after the spill and may be recovering" (Irons et al. 2000)

"We suggest Harlequins suffered population-level effects through 1992, but spill effects and regional ecologies cannot be separated to explain differences in abundance and productivity between oiled and unoiled areas" (Patten et al. 2000)

"Recovery of harlequin ducks has not occurred rapidly following this acute-phase mortality (in 1989) and there is evidence of persistent chronic effects of the oil spill" (Peterson 2001)

"Researchers still believe that continued hydrocarbon exposure is a potential contributing factor to (the) lack of recovery" (Exxon Valdez Oil Spill Trustee Council 2001)

"… Adult female winter survival was lower on oiled than unoiled areas during 1995–1998…" (Esler et al. 2002)

"Recovery of … harlequin ducks in the heavily oiled region of Knight Island has not occurred, with continuing oil exposure suspected as a factor" (Exxon Valdez Oil Spill Trustee Council 2002a)

"The 1989 *Exxon Valdez* oil spill took a terrible toll from which the Sound has not fully recovered" (National Wildlife Federation 2002)

"Taken together, the population census trends, survival measures and indications of exposure suggest that the harlequin duck has not recovered from the effects of the spill" (Exxon Valdez Oil Spill Trustee Council 2002b)

"Species such as … harlequin duck declined due to the oil spill and are not recovering" (National Wildlife Federation 2003)

"Female survival and CYP1A induction were similar between oiled and unoiled areas, suggesting that spill related injury was no longer occurring" (Esler et al. 2003)

"We conclude that there was no evidence of an impact on habitat occupancy by Harlequin Ducks" (Wiens et al. 2004)

"The lower proportions of females in oiled areas provided the only evidence for a possible lingering oiling effect. Demographic data interpreted in concert with other biological parameters leads us to conclude that harlequin duck populations are recovering from the *Exxon Valdez* oil spill" (Rosenberg et al. 2005)

"Our homogeneity of slopes test and regressions on summer and winter densities in oiled areas relative to unoiled areas of PWS did not show any evidence of a recovering population…. Higher levels of P4501A induction were found in oiled areas than unoiled areas for Harlequin Ducks …. These results are consistent with our trends showing … no recovery for Harlequin Ducks" (McKnight et al. 2006)

"… The most recent data from March 2005 irrefutably demonstrated that harlequin ducks continued to be exposed to lingering oil…" (Esler 2007)

"Harlequin ducks continued to be exposed to residual *Exxon Valdez* oil up to 20 years after the original spill" (Esler et al 2010)

"Direct effects of the oil spill on harlequin duck demography had largely abated by the winters of 200–2001 to 2002–2003 Esler and Iverson 2010)

statements are sometimes contradictory, but most center on possible spill effects on Harlequin body condition, reproductive performance, and/or population structure and dynamics. Early and persistent concerns were expressed about failed reproduction by birds occupying the spill area and about possible toxic effects of ingesting oiled mussels. In recent years, population declines have been linked to reduced overwintering survival of females, poor habitat quality, exposure to hydrocarbons (as assayed by cytochrome P450 (CYP1A); see Sect. 4.4.2.1), and/or reproductive failure. Poor reproduction and reduced survival in the spill area have also been linked with declining populations or a slower rate of population increase in oiled areas in relation to unoiled parts of PWS, although there is uncertainty about the effects of the spill on population dynamics. Fears also were raised about long population-recovery times, and some researchers concluded that there has been little recovery nearly two decades after the spill occurred. Esler et al. (2002) concluded that Harlequin Ducks had not yet recovered by 1998 and that full recovery was constrained by continuing exposure to residual oil. More recently, both McKnight et al. (2006) and the Trustees (Trustees 2009) concluded that Harlequin Ducks had not fully recovered from the effects of the spill by 2005 or 2009, respectively.

4.4   Evaluating the Evidence of Spill Effects

Before we assess the observations and data that bear on these concerns, it is useful to consider the ways in which an oil spill can affect seabirds. There are three pathways of possible spill effects: direct oiling of birds, oiling of the habitat, and consequences of the cleanup activities (Fig. 4.2). Each of these pathways involves a series of mechanisms that produce particular results. Collectively, these pathways influence the distribution (habitat occupancy), abundance, and reproductive performance and survivorship of populations (see Wiens 1995 for details).

Conceptual models such as Fig. 4.2 are useful because they require explicit identification of the potential pathways that link causes (on the left) with effects (on the right). It is immediately apparent, for example, that the variables that one measures to monitor spill effects (e.g., population size) may be affected by numerous pathways, both direct and indirect. Thus, although it might appear to be straightforward to detect a decreased population size of Harlequin Ducks in the spill area, it is much more difficult to determine which pathway(s), if any, caused the population decrease. One could argue, of course, that determining the precise pathway of a spill effect is not really necessary, because all of the pathways reflect consequences of the spill itself. The value of defining causal pathways, however, is that the *mechanisms* that link cause to effect are specified, providing possible solutions for mitigation. Documenting a mechanism helps to strengthen a causal explanation; conversely, the lack of a *bona fide* mechanism that can link an oil spill to a given consequence makes inferences about cause–effect relationships tenuous. Of course, other causal pathways that are unrelated to the spill event (not shown in Fig. 4.2)

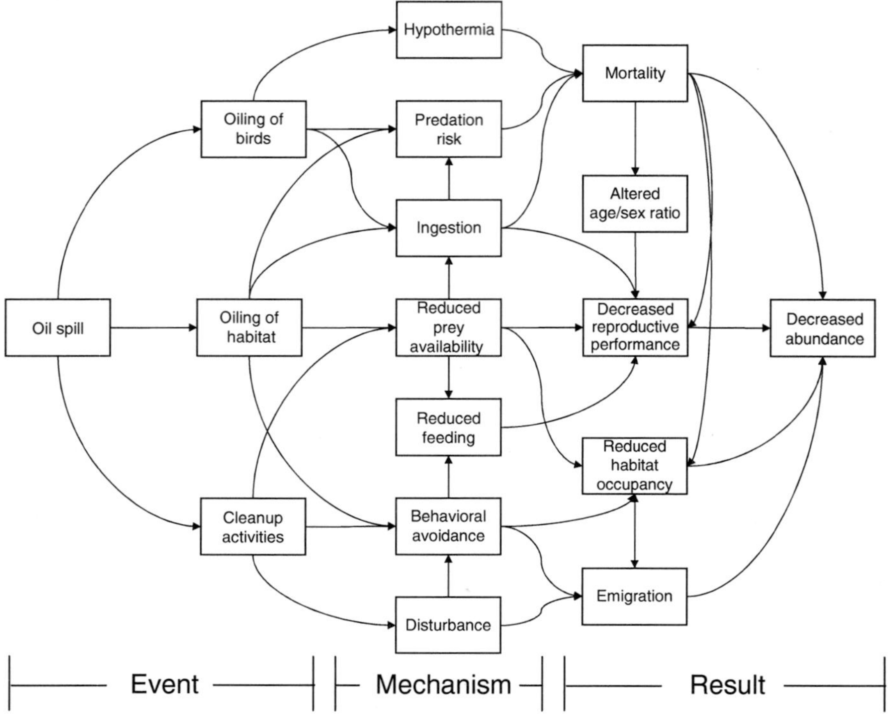

Fig. 4.2 Potential pathways through which an oil spill can affect birds

may lead to similar effects (e.g., decreased population size). This uncertainty explains why it is so important to develop and evaluate all reasonable hypotheses before attributing a given consequence to an oil spill (Harwell et al. 2010; Wiens 2008).

We can now evaluate how the *Exxon Valdez* oil spill affected Harlequin Ducks by examining the evidence in the context of the causal pathways shown in Fig. 4.2 and in terms of other potential pathways unrelated to the spill. We organize this evaluation around the three major consequences of oil spills – changes in abundance, in reproduction and survival, and in habitat occupancy and use.

4.4.1 Potential Effects on Harlequin Duck Distribution and Abundance

4.4.1.1 Direct Mortality

Direct mortality is the most obvious and clearest consequence of an oil spill. Overall, 212 oiled Harlequin Duck carcasses were retrieved from the spill area in

1989, 147 of them within PWS (Rosenberg and Petrula 1998). Models developed to generate total mortality estimates from carcass counts used a retrieval rate of 35% for PWS (Ecological Consulting Inc 1991); thus, ~423 Harlequin Ducks were estimated to have died in PWS as a result of direct exposure to oil (Patten et al. 2000; Rosenberg and Petrula 1998). Trustee researchers collected another 231 Harlequin Ducks for bioassays during 1989 and 1990, 132 of them within PWS. Overall, then, some 555 Harlequin Ducks died in PWS directly as a consequence of the spill and research associated with it. Based on late-winter population estimates (Table 4.2), this mortality would represent ~3–6% of the Harlequin Ducks present in PWS. Following Piatt and Ford (1996), Esler et al. (2002), used a 15% retrieval rate of carcasses. This would lead to a direct mortality of 980 birds in PWS, or roughly 7% of the wintering population of the area. These values were

Table 4.2 Population estimates and 95% confidence intervals for Harlequin Ducks in Prince William Sound, Alaska, 1972–2000

Year	March		July		August	
	Number	95% CI	Number	95% CI	Number	95% CI
1972[a]	12,480	3,325	3,607	2,038	–	–
1973[a]	15,831	5,528	–	–	18,218	27,281
1984/1985[b]	–	–	5,476	–	–	–
1989[c]	–	–	3,923	1,318	7,160	2,307
1990[c]	10,629	2,544	9,341	3,507	7,815	2,168
1991[c]	11,158	2,872	8,264	3,116	–	–
1993[d]	18,619	7,389	8,322	2,658	–	–
1994[e]	19,204	4,573	–	–	–	–
1996[e]	17,151	4,041	10,619	2,991	–	–
1998[f,g]	14,257	3,469	8,800	2,448	–	–
	14,621	3,486	8,853	2,425		
2000[g,h]	14,881	3,332	9,276	2,955	–	–
	14,876	3,288	9,191	2,868		
2004[g]	13,174	2,994	6,740	2,106	–	–
2005[g]	11,066	3,399	8,981	2,381	–	–

[a]Estimated from data of Dwyer et al. (unpublished), as reported in Klosiewski and Laing (1994) and Agler et al. (1999)
[b]From Irons et al. (1988); a census of birds along essentially the entire shoreline rather than a population estimate. In addition, no coastal-pelagic segments were done this year, so this number is for the nearshore component only. Segments were surveyed from mid-July to late August 1984 and from late May to late August 1985, although subsequent studies (Klosiewski and Laing 1994; Agler et al. 1994) have treated them as if they had been surveyed in July
[c]From Klosiewski and Laing (1994)
[d]From Agler et al. (1994)
[e]From Agler and Kendall (1997)
[f]From Lance et al. (1999); first numbers presented
[g]From McKnight et al. (2006); second numbers presented. No explanation is presented for the discrepancy in population estimates between the two reports
[h]From Stephensen et al. (2001); first numbers presented

used in the recent Trustee update of spill impacts on Harlequin Ducks in the immediate aftermath of the spill (Trustees 2006).

4.4.1.2   Prespill Abundance

Limited information is available on the abundance and distribution of Harlequin Ducks in PWS prior to the oil spill. Surveys conducted in 1972–1973 by U.S. Fish and Wildlife (USFWS) personnel [Dwyer, Isleib, Davenport, and Haddock (hereafter, Dwyer et al.), cited in Agler et al. 1999; Klosiewski and Laing 1994] indicated that perhaps 12,000–15,000 individuals were present in PWS in late winter (March), with numbers reduced to ~3,600 individuals in mid-summer (July; Table 4.2). Dwyer et al. estimated that over 18,000 individuals were present in August 1973 (Table 4.2), but that value had an extremely high error estimate (~50% higher than the population estimate), suggesting an abnormal distribution of birds and/or inadequate sampling, so its accuracy is questionable. Irons et al. (1988) censused the entire shoreline of PWS in the mid-1980s, recording 5,746 Harlequins in mid-summer (Table 4.2).

4.4.1.3   Seasonal Variation in Abundance

Although Harlequin Ducks occur in PWS throughout the year, densities are greatest in the fall and winter and are lowest during the breeding season, when nonresident birds depart for inland breeding areas and many resident females move into streams to breed (Table 4.2; Rosenberg and Petrula 1998). Numbers then increase in late summer and fall through recruitment and return of migrant individuals to molting and wintering areas. The Sound-wide surveys have estimated 11,000–19,000 individuals in late winter and (except for the aberrant estimate of Dwyer et al.) 5,000–10,000 birds in the summer.

4.4.1.4   Postspill Changes in Abundance and Distribution

Changes in abundance and distribution often represent the "bottom line" in evaluating the effects of an oil spill or, indeed, any environmental change. If effects on other components of the pathways shown in Fig. 4.2 are biologically important, they should ultimately translate into changes in abundance in the spill-affected area. In this section, we will describe in some detail the ways in which the effects of the *Exxon Valdez* spill on Harlequin Duck abundance have been measured and evaluated. Such a detailed examination is necessary to understand how the design of a study can affect its conclusions and to understand why such contrasting conclusions (e.g., Table 4.1) have been reached.

Comparisons to determine changes in Harlequin Duck abundance after the oil spill have been made in several ways and at several scales. Considered over PWS as a whole, estimates of overall abundance in late winter suggest a substantial reduction in numbers in 1990 from estimates made in the early 1970s by Dwyer et al. (a baseline design; Wiens and Parker 1995), with numbers matching or exceeding the prespill estimates by 1993 (Table 4.2). The summer surveys also suggest a reduction in overall abundance in July 1989 from the 1984 to 1985 prespill levels, again followed by a dramatic increase that exceeded prespill numbers by July 1990 (Table 4.2). In comparison with the 1972 summer surveys, Agler et al. (1999) reported a 107% overall increase in numbers in the 1989–1993 period as a whole; this was the largest proportional increase of any of the 25 taxa that they considered.

It is unclear whether these Sound-wide estimates indicate broad-scale, regional trends or more localized effects. The confidence intervals on the estimates are large, so these patterns are coarse at best. Further, because much of PWS was not directly affected by the *Exxon Valdez* spill, such estimates may provide little insight into actual spill-related effects. Other comparisons have sought to narrow the focus in several ways. One way is to compare abundance trends over time in eastern PWS, which was well away from the spill path, with those in western PWS, which was generally in the spill path [i.e., a before-after-control-impact (BACI) design; Irons et al. 2000; McDonald et al. 2000; Wiens and Parker 1995; Osenberg et al. 1994] (Fig. 4.1a). These two areas represent the main "oiled-control" comparison used in several of the postspill Harlequin Duck studies (e.g., Rosenberg and Petrula 1998; Crowley 1995; Patten et al. 1995, 2000; Patten 1993a, 1993b, 1994), although Rosenberg et al. (2005) later added a second study area within the general oil-affected area ("Southwestern PWS") and the control area ("Montague Island") to broaden the geographic scope of the study and to increase sample sizes. This comparison (Fig. 4.3) indicates no systematic changes in summer abundance over time in either area, although estimates of abundance in western PWS in 1984–1985 were somewhat greater than in eastern PWS. A separate analysis of seasonal variations in abundance revealed a strong correlation between the two areas ($r=0.664$; $n=9$; $P=0.026$). A comparison of estimated populations based on the full set of 1984 survey segments with estimates based on a subset of the 1984–1985 survey segments that was used in the subsequent analyses of Klosiewski and Laing (1994) suggests that these subsets systematically underestimated abundances in western PWS and overestimated abundances in eastern PWS (compare "1984" estimates with "1984–1990" estimates in Fig. 4.3).

These findings contrast with those of Patten et al. (1995, 2000), who surveyed areas of differing extent in western PWS in July–August 1991–1993 and compared their results with those obtained by Crowley (1994) in eastern PWS in 1991–1992 and by a separate field crew in 1993 (Table 4.3). Harlequin densities (birds/km of shoreline surveyed) were consistently lower in the western area than in the eastern area and decreased from 1991 to 1993, whereas densities in the eastern area were higher overall and fairly constant through time. The surveys were not randomly or systematically located, however, and the selection of survey sites apparently differed between eastern PWS (where Crowley focused attention on suitable breeding streams and shorelines) and western PWS (where surveys

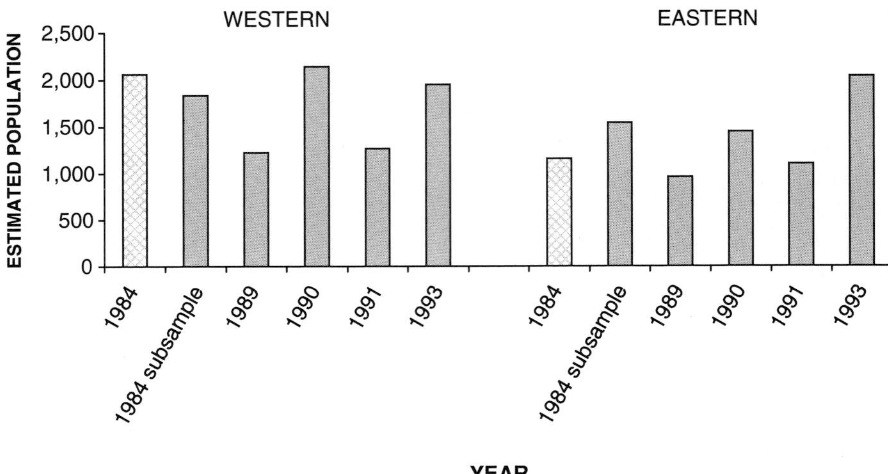

Fig. 4.3 Mid-summer abundances of Harlequin Ducks in eastern and western Prince William Sound (Fig. 4.1a). The 1984–1985 data (cross-hatched) are from Irons et al. (1988). The "1984 subsample" data are an estimate generated from a subset of the transects used by Irons et al. that were also surveyed by Klosiewski and Laing in 1989–1990. The 1989–1993 data are from Klosiewski and Laing (1994) and Agler et al. (1994)

Table 4.3 Abundances of Harlequin Ducks recorded in nearshore boat surveys in July–August 1991–1993 in western and eastern Prince William Sound

	Western PWS			Eastern PWS		
	1991	1992	1993	1991	1992	1993
Shoreline length surveyed (km)	537	2,276	1,296	700	410	620
Total Harlequins recorded	680	1,713	761	1,396	743	1,373
Linear density (individuals/km)	1.27	0.75	0.59	1.99	1.81	2.21

From Patten et al. (1995)

included a broad range of environmental conditions). It is unclear whether the differences among observers or survey-location selection contributed to the differences in Harlequin abundances between the regions or whether the change in survey extent in western PWS among years might have affected the likelihood of encountering birds. Because the surveys were conducted at different distances from shore, there may also have been confounding differences in detection probabilities.

Many of the same transects that were surveyed in Patten's studies were also surveyed in 1995–1997 by Rosenberg and Petrula (2001, 1998), who used more standardized procedures than did Patten. All survey transects in eastern PWS, however, were known to support high densities of Harlequin Ducks, whereas some known low-density areas were included among the survey transects in western PWS.

Thus, the likelihood of encountering Harlequins differed systematically between the areas for reasons that may have had nothing to do with oil-spill effects. Rosenberg and Petrula analyzed the survey data by evaluating how *trends* in abundance over the 3 years differed between the areas (a "parallelism" or "homogeneity of slopes" analysis; see Skalski et al. 2001; Skalski 1995) at three scales of analysis (individual transects, regions, and entire areas). The 1995–1997 trends were nonsignificantly positive for all scales in eastern PWS but were all negative in western PWS (but significantly so only at the regional scale; Table 4.4). Rosenberg and Petrula concluded that the abundance of Harlequin Ducks remained stable over the 3-year period in eastern PWS but decreased in western PWS, continuing the pattern reported by Patten.

Other studies have compared the oiled area of PWS with the entire unoiled area, which includes the northern and northwestern part of PWS as well as the eastern region used above (Fig. 4.1b). Klosiewski and Laing (1994) compared the results of the prespill, Sound-wide surveys of Dwyer et al. (in 1972–1973) and of Irons et al. (in 1984–1985) with their own postspill data (1989–1991) from a subsample of the earlier survey transect locations (i.e., a BACI design; Wiens and Parker 1995). Compared with the 1972–1973 surveys, Klosiewski and Laing found a significant decrease in the overall abundance of Harlequin Ducks in PWS in August [$P=0.09$; but note that this comparison involves the questionable August estimate of Dwyer et al. (Table 4.2)]. Comparisons with Irons et al. yielded only nonsignificant decreases and increases in overall population estimates between pre and postspill surveys in March ($P=0.17$) and July ($P=0.38$), respectively.

Klosiewski and Laing also evaluated whether abundance decreased or increased more in the oiled area over years than would have been expected from the yearly trend in abundance in the unoiled area (i.e., a pre/post pairs design; Wiens and Parker 1995). Population size decreased more strongly in the oiled part of PWS than in the unoiled part in July and August but not in March. In comparison with the 1984–1985 summer survey data of Irons et al., Klosiewski and Laing found a nonsignificant relative increase in Harlequin Duck numbers in the oiled part of PWS in 1989 ($P=0.88$) but significant relative decreases in abundances in the oiled region in both 1990 ($P=0.02$) and 1991 ($P=0.01$). In another comparison involving the same areas, Agler et al. (1994) and Agler and Kendall (1997) found a significant oiling effect on abundance trends in the oiled part of PWS in late winter but not in mid-summer (Fig. 4.4). In both periods, however, the increasing divergence over time between densities in unoiled and oiled areas was taken as evidence of continuing spill impacts (Agler and Kendall 1997). Using the same sort of parallelism analysis for 1989–2000 survey data from the same areas, Stephensen et al. (2001) concluded that the population trend in oiled areas for winter (March) surveys was consistent with a recovering population but that the summer (July) populations were not recovering.

McKnight et al. (2006) extended the Stephenson et al. (2001) data set through 2005 and concluded that the Harlequin Duck population in PWS still was not recovering (Fig. 4.4, Table 4.2). They based this conclusion on the observation that the slope of the regression of the population estimate against year for the oiled area was not significantly greater than that in the unoiled area (a "homogeneity-of-slopes" test).

Table 4.4 Population trend for Harlequin Ducks in oiled areas of western (WPWS) and unoiled areas of eastern (EPWS) Prince William Sound, Alaska, during the fall of 1995, 1996, and 1997

Study area	Number of slopes	Spatial scale	Weighted mean slope	Standard error	P[a]
EPWS	77	Transect	0.3114	0.2637	0.241
WPWS	54	Transect	−0.2022	0.3869	0.604
EPWS	15	Region	0.2476	0.1653	0.158
WPWS	21	Region	−0.6412	0.2580	0.023
EPWS	3	Study area	0.3258	0.3826	0.551
WPWS	3	Study area	−0.2422	0.4796	0.702

Analyses were conducted at the scales of individual transects, regions (combinations of transects in close proximity to one another and generally geographically distant from other regions), and study area (i.e., eastern and western PWS). Analyses were conducted on both sexes combined. Total numbers of individuals recorded: EPWS = 12,926, WPWS = 9,543. From Rosenberg and Petrula (1998)

[a]Probability of slope being significantly different from 0 (*t*-test)

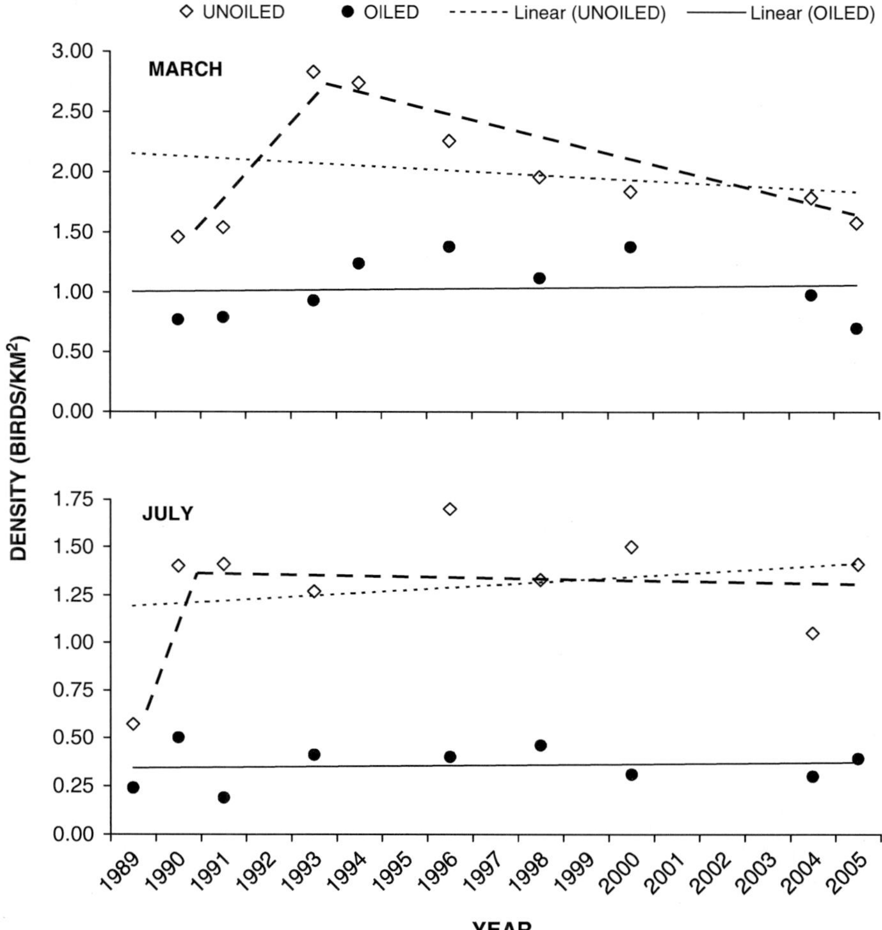

Fig. 4.4 Estimates of densities (birds/km²) of Harlequin Ducks in unoiled and oiled areas of Prince William Sound (Fig. 4.1b) in March and July 1990–1991 (Klosiewski and Laing 1994), 1993 (Agler et al. 1994), 1994 (Agler et al. 1995), 1996 (Agler and Kendall 1997), 1998 (Lance et al. 1999), 2000 (Stephensen et al. 2001), and 2004–2005 (McKnight et al. 2006). The regression lines are from McKnight et al. (2006)

A closer examination of the data plotted in Fig. 4.4, however, indicates that there are strong inflections in the data from the unoiled area for both March and July. To reflect the trends in the data more accurately, we have added multiple trendlines (fitted by eye) to Fig. 4.4. The revised trendlines indicate that the slope of the trendline in the oiled area is substantially greater than the (declining) trendline in the unoiled area in March. According to a homogeneity-of-slopes test, this indicates ongoing recovery since ~1993. The July data from 1990 onward follow similar, relatively stable trendlines. This parallelism of slopes meets one of the primary assumptions of a BACI analysis.

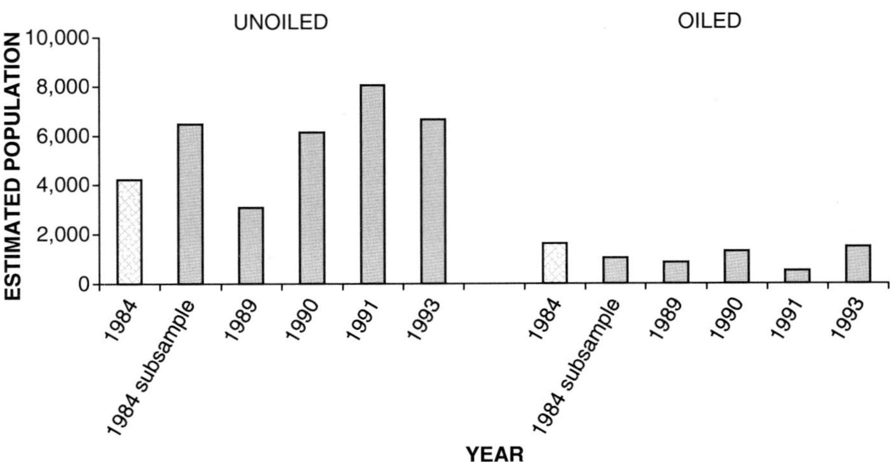

Fig. 4.5 Mid-summer abundances of Harlequin Ducks in oiled and unoiled areas of Prince William Sound (Fig. 4.1b). Data sources as in Fig. 4.3

Two additional analyses of the data sets of Klosiewski and Laing and Agler et al. provide some insight into the consequences of using different sets of transects or areas in comparisons. We reanalyzed postspill, summer abundance trends in the nearshore zone of the oiled and unoiled areas of PWS as they were defined by Klosiewski and Laing and by Agler et al. Comparing postspill estimates with the baseline of the entire Irons et al. data set (partitioned into oiled and unoiled regions) clearly indicates a postspill increase in abundance in the unoiled part of PWS but no postspill increase in the oiled region (Fig. 4.5). If we compare the prespill estimates generated only from the subset of Irons et al. transects that were subsequently surveyed by Klosiewski and Laing, however, there was a significant (53%) decline in abundance in the unoiled part of PWS in 1989 but a nonsignificant (18%) decline in the oiled area. The lack of correspondence in abundance trends between the oiled and unoiled regions is reflected in the poor correlation between the monthly estimates ($r=-0.193$; $n=9$; $P \gg 0.05$). These differences between the prespill abundance estimates of Irons et al. and of Klosiewski and Laing for both areas again suggests that the postspill transects surveyed by Klosiewski and Laing may not have been a representative subset of the original transects (see Fig. 4.3).

Rosenberg and Petrula (Alaska Department of Fish and Game; hereafter, ADFG; 1998) also compared trends during 1995–1997 with those derived from the Klosiewski and Laing and Agler and Kendall (USFWS) studies in 1989–1996. In a comparison that included the entire unoiled area surveyed by the USFWS, Rosenberg and Petrula found no significant temporal trends in either oiled or unoiled areas and no significant differences in the slopes of changes between the areas for any time periods in 1995–1997. When the comparison excluded those transects surveyed by the USFWS that lay outside of the eastern and western PWS areas defined by Rosenberg and Petrula, there was no significant trend in Harlequin abundances in eastern PWS for the ADFG data but a significant positive trend for the USFWS data; for western PWS, neither trend was significant, and slopes did not differ between eastern and western regions in either data set. Thus, the conclusions about relative trends differed depending on the analysis and geographic areas used in the comparisons. Rosenberg and Petrula attributed both the differences in results and the generally low power of tests involving the USFWS data to differences between the two studies in the allocation of survey effort within oiled and unoiled areas and to the inclusion in the USFWS data set of many areas containing low densities of Harlequin Ducks. One can also question the propriety of selecting for surveys only those areas in which large numbers of Harlequins occurred.

Despite these inconsistencies, it is clear that numbers of Harlequin Ducks were consistently much lower in the oiled than in the unoiled region of PWS, both before and after the oil spill. This difference suggests that these two regions differ fundamentally in their environmental suitability or attractiveness to Harlequin Ducks. The differences in numerical trends over time between the entire unoiled region (Figs. 4.4 and 4.5) and the eastern part of PWS (a subset of the unoiled region; Fig. 4.3) also suggest environmental differences among parts of the unoiled region. In particular, the increase in abundance of Harlequin Ducks in the unoiled area during the early 1990s appears to have occurred almost entirely in the glaciated fjords of northern and northwestern PWS, compromising the interpretation of the broad-scale patterns and the utility of the BACI design that included those fjords.

At a finer scale of analysis, we used a set of ten study bays that differed in the magnitude of oiling (from completely unoiled to some of the most heavily oiled bays in PWS) but were generally similar in other habitat characteristics (Day et al. 1995, 1997; see Wiens et al. 2001a). If we compare only postspill (1989–2001) densities in the set of four bays that were unoiled or lightly oiled (hereafter, unoiled) with those in the six moderately to heavily oiled bays (hereafter, oiled) (an impact-level-by-time design; Wiens and Parker 1995) (Fig. 4.6), it is apparent that densities were consistently greater in the unoiled bays than in the oiled bays in all years except 2001 and 2004 and that the year-to-year rate of increase in densities was greater in the oiled bays.

Because these same bays were also surveyed by Irons et al. as part of their 1984–1985 mid-summer surveys, we extended the time-line back to 1984 to compare the postspill trends with this prespill "baseline." This comparison changes the interpretation of the trends from that based only on the postspill surveys. In 1984–1985, densities of Harlequin Ducks were considerably lower in those bays that were later oiled than they were in the unoiled bays, suggesting that underlying habitat differences between the two sets of bays may have contributed to the apparent oiling effect. Moreover, densities in the subsequently oiled bays in 1984–1985 were only slightly greater than densities in those bays in 1989–1991, whereas densities in the unoiled bays declined substantially between 1984–1985 and 1990 (but not 1989 or 1991; Fig. 4.6).

Wiens et al. (2004) conducted statistical analyses of these pre/postspill comparisons by partitioning the analysis to determine whether changes in densities from prespill levels differed between unoiled and oiled bays [a pre/post pairs (BACI) design;

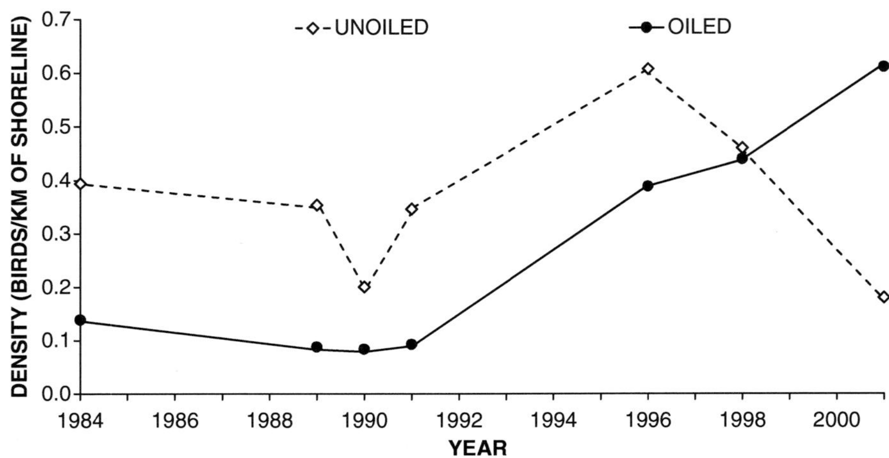

Fig. 4.6 Estimated population densities of Harlequin Ducks in mid-summer surveys of four unoiled/lightly oiled ("unoiled") and six moderately to heavily oiled ("oiled") bays in Prince William Sound. The 1984 data are from surveys conducted in the same bays by Irons et al. (1988); the 1989–1998 data are from Day et al. (1995, 1997, unpublished)

Wiens and Parker 1995]. Densities of Harlequin Ducks decreased in the oiled bays by 9.6, 5.4, and 1.1% relative to changes in the unoiled bays from the prespill baseline during 1989, 1991, and 1996, respectively, but increased by 25.3, 25.8, and 126.4% in 1990, 1998, and 2001, respectively. Of these increases, only the 1984–2001 increase was statistically significant. Nevertheless, this analysis revealed no clear indications of spill-related impacts on Harlequin densities in even some of the most heavily oiled bays in PWS shortly after the spill.

These results can be compared with those reported by Irons et al. (2000), who conducted a similar BACI analysis of Harlequin Duck abundances on a larger number of survey transects scattered throughout PWS. Irons et al. analyzed their data at three scales: individual transects (several of which might lie within a single bay), a medium scale combination of transects within a location (corresponding in size roughly to the bay scale used by Wiens et al.), and a coarse scale that grouped transects over the entire oiled or unoiled areas. At the medium scale, Irons et al. found a nonsignificant relative increase in Harlequin densities in oiled bays in 1989 from 1984 to 1985, followed by significant relative decreases in 1990 and 1991 and nonsignificant decreases in 1993, 1996, and 1998. Statistical power at this scale of analysis was moderate (0.47–0.64 at $\alpha=0.20$). The patterns were similar at the fine scale, although only the decrease in 1990 was significant. At the coarse scale (corresponding roughly to the scale used by Klosiewski and Laing and by Agler et al.), relative abundances in the oiled area in 1989, 1993, 1996, and 1998 were substantially greater than the prespill baseline and were significantly lower ($P<0.20$) only in 1991; however, power at this scale was generally low (0.23–0.62). Thus, whereas Wiens et al. found nonsignificant relative decreases from prespill levels in 1989, 1991, and 1996 and nonsignificant increases in 1990 and 1998,

Irons et al. reported nonsignificant relative increases in oiled areas in 1989, significant relative decreases in 1990 and 1991 (depending on the scale used), and nonsignificant decreases thereafter. Irons et al. concluded that there was "strong evidence of negative oil spill effects a few years after the spill" but that Harlequin Ducks "may be recovering." Wiens et al. (2001b) have commented on the sampling design and interpretations of this study (see also Irons et al. 2001).

4.4.1.5 Interpretation of Abundance and Distribution Changes

What can we conclude from these varied analyses of possible spill-related changes in the abundance of Harlequin Ducks following the *Exxon Valdez* oil spill? The comparison of eastern PWS with western PWS revealed no significant differences in trends in postspill abundance between the regions, although there were indications that postspill increases in the western region were less than those in the eastern region and may actually have been decreases. Other comparisons indicated that (1) Harlequin Ducks were more abundant in the unoiled than in the subsequently oiled regions of PWS, even before the spill occurred; (2) abundances continued to be lower in oiled than in unoiled areas several years after the spill; (3) some studies recorded proportionately greater decreases in 1989 relative to prespill abundances in oiled areas, whereas other studies found greater decreases in unoiled areas; (4) several studies found clear evidence of reduced densities in oiled areas in 1990 and 1991 relative to prespill baselines; and (5) abundances increased during the mid- to late 1990s to levels equal to or greater than the prespill estimates in both oiled and unoiled areas, although different studies recorded greater postspill increases in either oiled or unoiled areas. In our view, these findings are consistent with a limited negative effect of the spill on Harlequin Duck distribution and abundance, followed by apparent recovery within a few years.

Several factors confound these conclusions, however. First, the estimated direct mortality of Harlequins from the spill was small, so it would have been difficult to detect the loss of a few hundred birds out of thousands over the spill area, much less over the entire Sound. (Of course, abundance may be influenced by factors other than direct mortality (Fig. 4.2); we consider potential effects on reproductive performance and habitat occupancy below.) If, however, ducks were to emigrate from the spill area in response to the oil or cleanup activities, abundance in the oil-affected area would be reduced, especially if individuals returned to the affected areas only after several years. Although there is evidence from banding studies that many adults exhibit strong site fidelity to molting or wintering locations but subadults less so (Iverson et al. 2004; Robertson et al. 2000; Holland-Bartels et al. 1998), the extent of such site fidelity in the population as a whole is unknown. Mark-recapture studies are likely to record those individuals that return to the study sites but not those individuals that have moved elsewhere. Further, genetic studies of Harlequin Ducks in PWS found no evidence of population structuring or subdivision (Lanctot et al. 1999), suggesting that movement of individuals is substantial enough to prevent genetic differentiation among localized populations.

Second, uncertainties about the baseline data available for comparison further complicate interpretations. The 1972–1973 data indicate abundance 16–17 years before the spill occurred and would be a suitable baseline for comparisons *only* if nothing had changed over the interim (i.e., the system as a whole is in steady-state equilibrium). This assumption is clearly violated (Wiens 1995, 1996). There have been several *El Niño* events, at least two marine regime shifts have affected the region (Hare and Mantua 2000; Peterson and Schwing 2003), sea-surface temperatures have undergone decadal-scale shifts (Francis et al. 1998), and the prey base for seabirds has changed dramatically since the 1970s (Golet et al. 2002; Agler et al. 1999; Piatt and Anderson 1996). The 1984–1985 data are more recent and therefore are a more reliable baseline. Except in BACI analyses, the use of prespill data to evaluate spill effects also assumes steady-state equilibrium in the absence of the oil spill.

Third, the abundance comparisons described above encompass quite different spatial scales and survey designs. The Sound-wide estimates show that overall numbers of Harlequin Ducks vary among seasons and years, but it is not clear whether the increases in numbers reported in the 1990s represent real population increases or simply reflect increased survey effort (itself a consequence of the oil spill). More to the point, such broad-scale analyses cannot really reveal spill impacts. The comparisons among broadly defined oiled and unoiled regions of PWS assume that these regions were adequately sampled and that there were no systematic spatial variations in abundances within the designated areas. The underlying assumption is that, prior to the spill, Harlequins "were abundant and distributed throughout the entire Prince William Sound" (Patten 1994). The differences in population trends between the overall unoiled region and eastern PWS (a subset of the unoiled region), as well as the differences between oiled and unoiled regions evident in the data of Irons et al., clearly show that these assumptions are almost certainly false. Similarly, the comparison between eastern and western portions of PWS assumes that these areas are ecologically similar except for the presence of oil, so eastern PWS can serve as a "control" for the western "treatment." As we will show below, this assumption is also invalid. The comparisons between eastern and western PWS may be further compromised by the problems in survey design (i.e., nonrandom designation of survey locations, differences in survey-site selection criteria and survey intensity, among-observer variation) noted above.

Fourth, interpretation of relative population trends over time between unoiled and oiled areas (e.g., Fig. 4.4) may also be problematic. As we have noted, the abundance of a species might differ between a set of oiled sites and another set of unoiled (reference) sites, independently of any oiling effects. Abundances might also vary among years, but if there is no oiling effect, one might expect the slopes of the overall trends in abundance over several years in the two areas to be the same. This is the "parallelism" or "homogeneity-of-slopes" approach advocated by McKnight et al. (2006), Skalski et al. (2001), Lance et al. (2001), and Stephensen et al. (2001). The expectation of homogeneity of slopes holds if both areas experience a constant arithmetic increase (i.e., both populations are incremented annually by the same number of individuals) or if populations show a constant proportionate increase (i.e., populations are incremented annually by the same percentage, as might occur if there were a region-wide population increase) and the data are expressed as log (density). A lower rate of increase in the oiled area relative to that in the unoiled area (i.e., diverging slopes) would therefore indicate continuing impact, whereas a more rapid increase in the oiled area (converging slopes) would suggest recovery from an initial impact. All of these interpretations (as well as the BACI analyses of Murphy et al. (1997), Irons et al. (2000), and Wiens et al. (2004)) depend on the assumption that, in the absence of an oil spill, the rates and patterns of change would be the same in both areas (Wiens et al. 2001b; Stephensen et al. 2001). This assumption is tenuous at best. For example, if the capacity of habitat to support Harlequin Ducks is inherently poorer in those areas that happened to end up in the spill path, rates of population change almost certainly will be substantially lower than those in the unoiled area, even in the absence of a spill. Hence, interpretation of differences in the slopes of population trends is not as straightforward as it might seem (see Hatch 2003). Moreover, the test itself requires that the data be plotted as simple linear regressions; the revised trendlines in Fig. 4.4 cast doubt on the efficacy of this approach.

Finally, the fine-scale studies of individual bays that differed in the magnitude of oiling from the spill may impose some degree of control over environmental variation among bays through the selection of study sites, and oiling can be treated as a quantitative rather than a categorical variable, facilitating more rigorous statistical analyses (see below). Nonetheless, there may still be environmental differences among bays that can obscure the effects of oiling (Wiens et al. 2001a; Day et al. 1997), and the sample size is inherently small because the number of possible sample units (bays) is limited. There is inevitably a trade-off between the intensity and quantitative rigor of a study design and the spatial scope, sample size, and statistical power of the study (Wiens and Parker 1995). The emerging message is that how a comparison is structured affects both the findings and the confidence one can place in them. Unfortunately, it is not obvious that any single approach or scale is necessarily "best," although some are probably better than others and some are clearly inadequate.

4.4.2 Potential Effects on Harlequin Duck Reproduction and Survival

4.4.2.1 Reproduction

Shortly after the *Exxon Valdez* spill, concerns were voiced about effects on the reproductive performance of Harlequin Ducks (Table 4.1). Initial observations suggested that productivity was extremely low in the spill area, and it was soon concluded that Harlequins in the spill zone had suffered "a massive reproductive failure" (Patten 1993a). This conclusion was promptly repeated in the media (e.g., Chadwick 1993; Edgar 1993; Pain 1993), although Wheelwright (1994) later considered it more critically. What was the evidence for such a conclusion?

Harlequin nests are difficult to find, so most assays of reproductive productivity in the spill area have been based on counts of Harlequin broods encountered after females bring their chicks to estuaries, sheltered bays, or shorelines. Prespill observations were opportunistic, depending on where investigators were working and whether they bothered to record broods or could identify them. Older broods are hard to distinguish from the large number of molting birds present in PWS in late summer, so there is some uncertainty about brood counts at that time (see Rosenberg and Petrula 1998). These records are scattered widely over PWS, but broods were reported in several locations within the area that was subsequently oiled by the spill (e.g., Oakley and Kuletz 1979; but later repudiated by Oakley and Kuletz, personal communication). Because the prespill observations were not made systematically, however, it is not possible to use them to derive quantitative estimates of reproductive activity or productivity in different parts of PWS; however, they do indicate how widely Harlequins nested at that time.

Following the oil spill, efforts were made to conduct more intensive studies of breeding Harlequin Ducks in both the eastern and western portions of PWS (Fig. 4.1a). Based on his own surveys and those of Crowley (1994), Patten (1994; Patten et al. 1995) calculated that Harlequin Duck production in eastern (unoiled) PWS was 2.9 broods/100 km of shoreline in 1991, 0.9 broods/100 km in 1992, and 1.8 broods/100 km in 1993. In contrast, production in western (oiled) PWS was 0.7 broods/100 km in 1991, 0.1 broods/100 km in 1992, and 0.2 broods/100 km in 1993[1]. As noted above, the survey extent in western PWS was broadened considerably in 1992 and 1993 (Table 4.3) and included a greater amount of unsuitable habitat than in the 1991 surveys. In addition, a cold, wet spring may have depressed production in both regions in 1992 (Patten 1993a; see also Ganter and Boyd 2000). Although they did not report their observations in the same way, Rosenberg and Petrula (1998) recorded 10, 14, and 12 broods in eastern PWS and none in western PWS in 1995, 1996, and 1997, respectively.

During our studies in Prince William Sound, we also recorded observations of Harlequin Duck broods. Between 1989 and 1993, we saw broods in both unoiled and lightly oiled bays and in areas subjected to moderate or heavy oiling in western PWS. We conducted only limited surveys in eastern PWS in 1996 and 1998 (where we also observed broods), so we cannot relate our observations to the east–west comparisons of Patten and Crowley. It is apparent, however, that some successful reproduction was occurring in the oil-spill area after the spill, even in the spill year (1989), when we saw broods not only in unoiled bays but also in the moderately oiled Bay of Isles; the consistent factor in all of these observations was their association with bays that had substantial salmon-spawning streams. Some of the concerns about complete reproductive failure expressed in Table 4.1, therefore, appear to be overstated. In fact, Rosenberg and Petrula (1998, 1999) questioned whether brood counts can provide a reasonable index of reproductive success in PWS, and

---

[1] Apparently, Patten neglected to record where his surveys began and ended; therefore, he had to reconstruct the shoreline distances surveyed several years after the fact, casting doubt on the reliability of the results.

by 1999 the Trustees declared that "conclusions of reproductive failure based on lack of broods in the oiled area do not now seem warranted" (Trustees 1999).

On balance, it does seem clear that productivity by Harlequin Ducks was greater during the postspill study years in eastern PWS than it was in western PWS (cf. Crowley 1994). The critical question is, Why? Patten (1993a, 1993b, 1994) offered two mechanistic hypotheses to account for the lower reproduction by Harlequins in the spill area: disturbance and chronic exposure to hydrocarbons.

Harlequin Ducks are sensitive to disturbance of many types (Goudie 2006; Goudie and Jones 2004; Perfito et al. 2002; Clarkson 1992; Wallen 1987; Dzinbal 1982), and the intensive cleanup activities during the summer of 1989 might have caused birds to leave areas that otherwise would have been occupied for breeding. Because cleanup activities were reduced in 1990 and minimal in 1991, Patten reasoned that one might expect to see an increase in productivity if disturbance had caused the low productivity in western PWS in 1989. Patten's observations suggested that productivity was in fact even lower in 1990 and 1991 than in 1989, so he rejected this hypothesis. Of course, if birds that left the area because of disturbance in 1989 bred elsewhere that year and then returned to those new breeding areas in subsequent years, one would not see the anticipated rebound in productivity once disturbance was reduced. Consequently, the Patten hypothesis cannot be completely discounted.

Patten's hydrocarbon hypothesis was more involved. He argued as follows (Fig. 4.7). Many beds of blue mussels (*Mytilus trossulus*) were oiled during the spill. In some cases, oil became trapped beneath the mussels' byssal-thread mats in anoxic sediments, where the oil retained the some toxic components. Petroleum hydrocarbons also became concentrated in mussel tissues from hydrocarbons in the water and the sediments. Patten also noted that Harlequin Ducks feed on blue mussels, consuming the entire mussel as well as the attached byssal threads (and some associated sediments), all of which could contain high levels of petroleum hydrocarbons. The birds could then accumulate elevated levels of these contaminants in their tissues, causing reproductive failure. Patten suggested that, because some mussel beds in the spill zone continued to retain oil for years after the spill, these effects would be chronic, contributing to continuing population declines in the spill area and eventually resulting in local extinction. This hypothesis has subsequently been extended to apply to a large suite of species (Peterson 2001); in a somewhat modified form, it is the basis for conclusions that Harlequin Ducks in PWS have not yet fully recovered from the effects of the oil spill (Trustees 2009).

Patten supported this scenario both with the results of the brood surveys discussed above and by noting that reproduction in the Naked Island group (which was in the spill path but only lightly oiled) declined from 156 juveniles in 11 broods reported in 1978 (Oakley and Kuletz 1979) to no broods in 1990–1992. (There is some doubt, however, that the prespill observations were actually of broods; Rosenberg and Petrula 1998; 1999; Oakley, USGS-BRD, personal communication.) Further, three Harlequins collected in western PWS had higher tissue hydrocarbon levels than did birds collected from several unoiled areas well away from the spill (and from PWS). None of the 104 sea ducks of several species examined for histopathology,

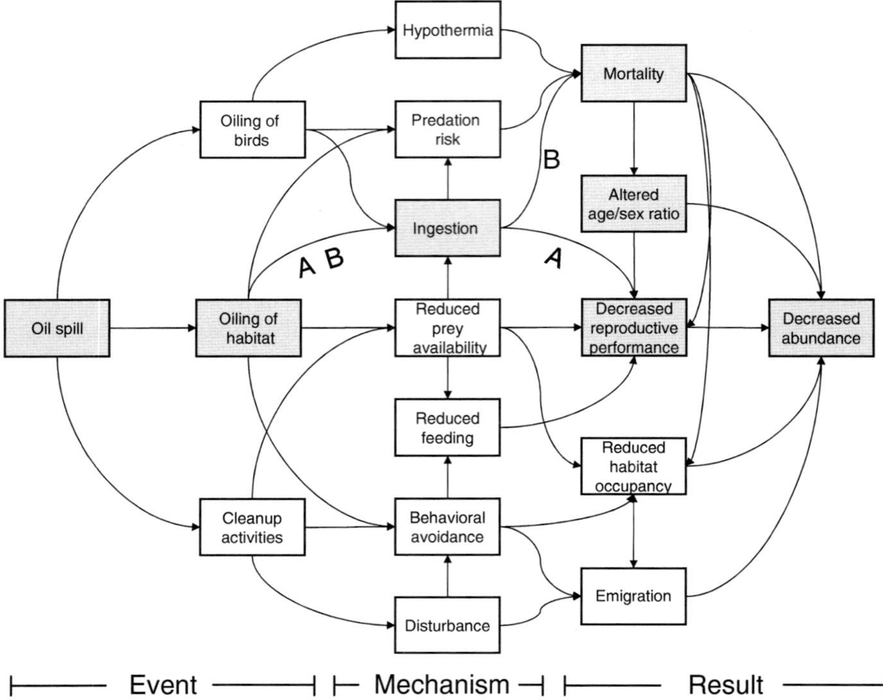

Fig. 4.7 Pathways through which different investigators have hypothesized that the *Exxon Valdez* oil spill affected Harlequin Ducks. Pathway A, advanced by Patten (1993a, 1993b, 1994) attempted to link reproductive failure and "local extinction" to ingestion of contaminated mussels. Pathway B, advanced by Esler et al. (2000a, 2002) and Esler (2008) attempted to link mortality of overwintering females to exposure to *Exxon Valdez* oil residues, as evidenced by elevated levels of Cytochrome P450 1A in the livers of ducks in the oiled region (Fig. 4.1a)

however, showed evidence of spill effects (Patten et al. 2000). In 1992, Patten and his colleagues found that 32 of 121 mussel beds examined actually contained significant amounts of subsurface oil (NOAA/NMFS investigators identified an additional 18 beach segments containing oiled mussel beds; Patten et al. 2000). Patten also observed that molting Harlequins frequently occurred on oiled mussel beds. Finally, he examined the prey items in the birds that were collected and found that some birds had ingested contaminated mussels (Patten et al. 1995, 2000).

Because Patten postulated a series of explicit linkages between the oil spill and the observations of low reproduction in the spill area (Fig. 4.7), we can examine and evaluate the evidence bearing on each link. First, large areas of mussel beds were indeed oiled by the *Exxon Valdez* spill, some of them quite heavily. Shortly after the spill, a decision was made that, to protect the integrity of the mussel bed communities, beds on fine-sediment substrates should not be cleaned (Babcock et al. 1996). Because some of these beds retained oil in the underlying fine sediment and the dense byssal mats hindered natural cleaning processes, less weathered oil persisted below the mats. Overall, however, the number and extent of such chronically oiled mussel beds that remained in PWS 1–2 years after the spill was small. After extensive searches, Patten and his colleagues (Patten 1994) found "over 50" oiled mussel beds, most of them located in protected, low-energy sites with fine-grained sediments (Babcock et al. 1996); such sediments form <4% of the shoreline of the spill area (Boehm et al. 1996; Page et al. 1995). Moreover, most of these beds were small, and the distribution of oil within such beds was patchy (Babcock et al. 1996; Harris et al. 1996). Based on surveys conducted in 1995, Carls et al. (2001) reported that total PAH concentrations in sediments at 27 of 34 sites in the most heavily oiled portions of the most heavily oiled locations in PWS (not necessarily mussel beds) were more than twice background levels. In two heavily oiled locations, however, Boehm et al. (1996) estimated that <3% of the mussels present in 1993 actually occurred on contaminated sediments and that, across all oiled sediments in PWS, the percentage of mussels associated with oiled shorelines was probably much lower.

The potential for mussels to be contaminated with *Exxon Valdez* oil residues is linked closely to the amount and distribution of these residues that remain in beach and intertidal sediments in PWS. Both Hayes and Michel (1999) and Carls et al. (2004) concluded that the geomorphic structure of beaches rather than the occurrence of mussel beds was responsible for oil being retained in some sediments. Surveys at a broad range of previously oiled sites in PWS recorded PAH concentrations at prespill background levels by 1990 or 1991 (O'Clair et al. 1996; Wolfe et al. 1996). In 1999, the Trustees reported that oil remained in "at least 30 sites" in PWS. As noted previously, Short et al. (2004) reported that their 2001 survey projected that a cumulative area of ~11.3 ha of shoreline in Prince William Sound was still contaminated by surface or subsurface oil residues. Most of this survey was conducted on cobble beaches lacking mussel beds, and most sampling stations were located in areas that were heavily oiled in 1989 and that had substantial oil remaining in 1990–1993. Although weathered oil persists in sequestered forms at beaches with boulder–cobble surface "armor" as long as 18 years after the spill (Michel et al. 2006; Taylor and Reimer 2008), what residues do persist are highly weathered and are found in scattered patches at limited locations that in turn are scattered along the 783 km of shoreline that had been oiled in 1989 (Boehm et al. 2008; Page et al. 2008; Short et al. 2007). Overall, it appears that the likelihood of a Harlequin actually encountering contaminated mussel beds within the entire spill-affected area would have been small by the mid- to late 1990s.

The second component of Patten's scenario is the bioaccumulation of hydrocarbons in the tissues of mussels occupying oiled beds. PAH levels were indeed high in mussels in heavily oiled areas immediately after the spill but decreased by several orders of magnitude over the next 2 years (Short and Harris 1996; Boehm et al. 1995; Shigenaka and Henry 1993). In most cases, PAH concentrations in mussel tissues within a few years of the spill were well below those known to affect the growth of mussels (Widdows et al. 1995) and PAH residues in mussel populations over the spill area of PWS were back to background levels by 1998–2000 (Neff et al. 2006; Integral Consulting 2006; Page et al. 2005; Boehm et al. 2004; Carls et al. 2004). Thus, the probability that a foraging Harlequin would encounter a contaminated mussel bed was low, and the probability that a foraging Harlequin would encounter mussels with high tissue concentrations of PAHs within those contaminated beds would be lower still, even if it were to feed predominately in oiled beds.

This series of linked effects assumes that Harlequin Ducks actually consume large quantities of blue mussels. Extensive diet data from PWS are not available, but Patten's own studies (1994) indicated that blue mussels constituted only 12% of the food volume (apparently including the indigestible shells) and 8% of the prey individuals of the Harlequins collected in PWS. Studies elsewhere on the West Coast (e.g., Fischer and Griffin 2000; Robertson and Goudie 1999; Gaines and Fitzner, 1997; Dzinbal and Jarvis 1984; Vermeer 1983; Kenyon 1961; Murie 1959) have found similarly low proportions of blue mussels in Harlequin diets. Harlequins also select only small (<5 mm) mussels, which were largely absent from heavily oiled mussel beds in PWS for several years after 1989 (Boehm et al. 1996). Small (young) mussels are generally present in a mussel bed for only a few years after the establishment of a bed, which usually follows a single settlement event. Such young mussels might have carried significant PAH contamination immediately after the spill in 1989 and 1990, but not in subsequent years, when the amount of oil in the environment was substantially less. In addition, small mussels occur primarily on macroalgae on the lower shore and shallow subtidal zone, where little or no residual *Exxon Valdez* oil persisted. Direct ingestion of relatively unweathered oil would occur primarily via the byssal threads, which would occur only with the larger mussels that Harlequins do not eat. Finally, Patten et al. (2000) reported that hydrocarbon-contaminated food was present in only 3 (4%) of 75 Harlequins collected for diet analysis in PWS in 1989–1990, when there was considerable surface oil on the shoreline. None of these individuals contained mussels among the diet items. Given these findings, it seems unlikely that, more than a decade later with far less oil present (and that primarily below the boulder–cobble surface), individuals could consume sufficient quantities of heavily contaminated mussels to produce the hypothesized physiological effects, much less the population-wide effects (Wiens 2007).

The next component in the hypothesized linkage of spill-related effects on Harlequin reproduction is the uptake of contaminants by the birds. Patten collected 104 ducks for analysis of hydrocarbon concentrations in organs and tissues and concluded that there were indications of elevated exposure to hydrocarbons in birds from western PWS relative to those from eastern PWS in 1989 (Patten 1993a). However, the analyses necessary to support this conclusion were never completed (Wheelwright 1994). Bence and Burns (1995) conducted a fingerprint analysis of the PAH compositions reported for the stomach contents of the Harlequins collected by Patten in 1989–1990. Of 18 birds analyzed in 1989, two had detectable *Exxon Valdez* hydrocarbon residues; of 56 birds analyzed in 1990, only one had detectable *Exxon Valdez* residues.

More recent studies (Esler et al. 2010, 2002; Esler 2008; Bodkin et al. 2002a, 2003; Trust et al. 2000) have suggested that there may indeed be differences in hydrocarbon exposure between birds in oiled and unoiled parts of western PWS. This conclusion rests on the use of a biomarker, the cytochrome P450 mixed-function oxygenase

(CYP1A) system, which is induced (its activity is increased) in the liver and other tissues by exposure to a variety of natural and anthropogenic chemicals in the environment (Stegeman and Hahn 1994). Trust et al. (2000) and Esler et al. (2002 2010) reported that liver samples collected in 1998 and 2000 and 2005–2009 from Harlequin Ducks in the Knight Island area (oiled) had higher levels of CYP1A (as measured by ethoxyresorufin-*o*-deethylase (EROD) activity) than did those from birds collected at an unoiled reference site on northwestern Montague Island. They concluded that the differences were likely due to exposure to residual oil from the *Exxon Valdez* spill and speculated that the biochemical and physiological changes in birds chronically exposed to oil were constraining the recovery of Harlequin populations. Bodkin et al. (2002a) confirmed the differences between areas in 2000 and offered much the same interpretation. Elevated levels of CYP1A in sea otters (*Enhydra lutris*) and nearshore fishes in oiled areas of PWS have also been related to exposure to residual *Exxon Valdez* hydrocarbons (Bodkin and Ballachey 2004; Bodkin et al. 2003; Jewett et al. 2002; but see Boehm et al. 2003; Jewett et al. 2003).

The CYP1A story is not as strong as would first appear, however, for several reasons. First, all of the reference (unoiled) samples were collected from birds on Montague Island. Although Montague Island does indeed lie largely (but not entirely) outside of the spill path, it also differs environmentally and oceanographically from most of the spill zone. Montague Island was uplifted during the 1964 Alaskan earthquake and now has a broad, shallow nearshore area. Nearshore currents and circulation patterns therefore differ from those in the deeper waters that characterize most of the spill area, affecting the exposure of birds to hydrocarbons.

Montague Island has served as the reference area not only for the CYP1A studies but for evaluations of overwinter survival of Harlequin Ducks in oiled vs. unoiled areas of PWS as well (see Sect. 4.4.2.2). We therefore used recent data on physical and biological variables from survey segments to compare shoreline habitats on Montague Island with those in the spill-affected area (Knight Island group and adjacent areas) (Parker et al., unpublished data). The statistical analyses (MANOVA and principal components analysis) indicated that there was essentially no overlap in habitat characteristics between the two areas for either the physical variables ($P<0.001$) or the biological variables ($P<0.001$) (Fig. 4.8). Montague Island does not seem to be an appropriate reference area to evaluate spill effects on Harlequin Ducks in PWS.

Second, although mean CYP1A levels (as measured by EROD activity) were more than 2 times higher in Harlequin Ducks from the Knight Island area than from Montague Island in 1998 and 2000, there was less than a twofold difference in levels between the areas in ducks collected in 2001, 2002, and 2005 (Esler 2008; Esler et al. 2002; Trust et al. 2000), although the differences may have been greater in subsequent years (Esler et al. 2010). Overall, levels of CYP1A in both oiled and reference locations were highly variable among years and among individuals within years. Mean EROD activity in samples from the reference area over the five sampling years between 1998 and 2005 ranged from 70.7 ± 21.5 to 218.3 ± 59.9 pmol/min/mg protein, whereas those in the oiled area ranged from 40.2 ± 18.4 to 776.6 ± 193.0 pmol/min/mg protein during the same time period. (Esler et al. (2010)

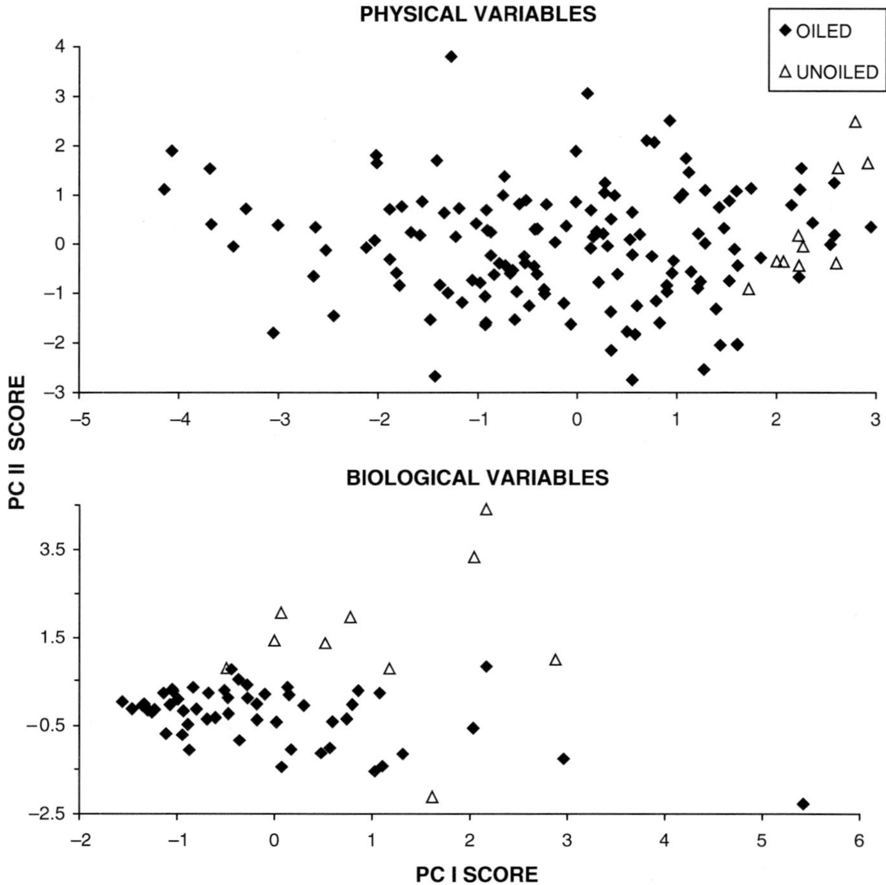

Fig. 4.8 Principal Component Analysis results for differences in physical habitats (top) and biological habitats (bottom) between shoreline segments on Montague Island (unoiled reference area) and shoreline segments on the Knight Island Group and the northwestern part of the spill-affected area (oiled area). The positions of oiled and unoiled symbols indicate that shoreline and nearshore habitats differ significantly between oiled and unoiled areas. Had habitat features been similar, the oiled and unoiled symbols would be randomly placed and thoroughly mixed. For the physical habitats, important features of Principal Component I are percent bedrock shoreline, islands/km shoreline, and percent boulder/cobble shoreline; important features for Principal Component II are percent bedrock/rubble shoreline, shallowness index, and islets/km shoreline. For the biological habitats, important features of Principal Component I are salmon runs/km of shoreline and percent of shoreline with seagrass; important features for Principal Component II are percent of shoreline with mussels and percent of shoreline with *Fucus* cover

do not report EROD levels for the 2006–2007 or 2009 samples, precluding an assessment of variation). Esler concluded that, however the samples and data were analyzed, CYP1A induction was greater in oiled than in unoiled areas in 1998 and 2000, that values converged in 2001 and 2002, and that CYP1A levels were again higher in oiled than in unoiled areas in 2005, 2006–2007, and 2009 (Esler et al. 2010; Esler 2008). Esler et al. (2010) concluded that these results "provide strong evidence of CYP1A induction in harlequin ducks from oiled areas, which we conclude is due to exposure to residual *Exxon Valdez* oil, and indicates that harlequin ducks remained at risk of potential deleterious consequences of that exposure."

Esler clearly related these results to the exposure of Harlequin Ducks to residual *Exxon Valdez* oil. However, CYP1A is induced by exposure of marine animals to bioavailable forms of a wide variety of compounds, such as PAH's, PCBs, dibenzo-p-dioxins, some chlorinated hydrocarbon pesticides, and a large array of other poorly characterized chemicals dissolved in the water column, in food, or in sediments (Lee and Anderson 2005; Hahn 2002). PAHs also differ in their ability to induce CYP1A activity; in general, the PAH's that are most abundant in petroleum (petrogenic PAH's) are weaker inducers than the PAH's that are more abundant in combustion (pyrogenic) hydrocarbon mixtures (Neff et al. 2005; Neff 2002). There are many sources of both petrogenic and pyrogenic hydrocarbons in PWS other than the *Exxon Valdez* – fuel leaks at sites of former or current human activity, fuel spills associated with boating activity, eroding shale and oil seeps southeast of PWS, oil released by the 1964 Alaska earthquake, natural peat deposits, engine exhausts from boats and machinery, air-borne particles from industrial activity or forest fires, or even bonfires on a beach (Bence et al. 2007; Neff et al. 2006; Page et al. 1999, 2002b, 2006; Lee and Anderson 2005; Huggett et al. 2003; Boehm et al. 2003; Mudge 2002; Kvenvolden et al. 1995).

The level of CYP1A activity induced by exposure to PAH's or other inducing chemicals is also affected by factors that vary among individual birds and among seasons. Differences among individuals in their proximity to areas of human activity can affect both their level of stress and their exposure to PAHs associated with those activities. The mobilization of the mixed-function oxygenase system is increased during the development of reproductive organs (Knight and Walker 1982; reviewed in Rattner et al. 1989), so individuals of different reproductive status or collected at different seasons may differ in the level of CYP1A induction, even if they have similar exposure to PAH's. Dietary differences of individuals among areas may also affect CYP1A induction (Oris, personal communication).Thus, although exposure to some PAH's does indeed elevate levels of CYP1A in Harlequin Ducks, the variations among individuals, years, and areas in CYP1A levels may reflect the influences of a variety of factors (Forbes et al. 2006; Lee and Anderson 2005). The difference in CYP1A levels between reference and oiled areas alone cannot be used to conclude that Harlequin Ducks from the oiled area of PWS continue to be exposed specifically to residual PAH's from the *Exxon Valdez* spill, as has been suggested by some (e.g., Esler et al. 2002, 2010; Esler 2008; Irons et al. 2000).

Finally, variation aside, we must ask what an elevated level of CYP1A in one area relative to another area really means, biologically. Although CYP1A levels in birds from Knight Island were at times more than double those from the Montague Island reference area, one can question whether statistically significant differences in EROD activity are also biologically significant (Oris and Roberts 2007). CYP1A indicates a normal response of organisms to exposure to chemicals in the environment, but it is not in itself an indicator of harmful individual or population-level effects (Forbes et al. 2006; Neff 1985). A linkage between PAH-induced CYP1A activity in birds and toxicological or physiological effects that would impair reproduction has not been established (Esler et al. 2002, 2010; Leighton 1993). The laboratory experiments conducted by Bodkin et al. (2002a) indicated that food consumption rates were significantly greater in Harlequins exposed to oil than in control birds, but metabolic rate, daily energy expenditure, and feeding behavior did not differ significantly between groups. Moreover, although the doses of crude oil administered to birds in these experiments resulted in a 7- to 11-fold induction of hepatic EROD activity, there was no dose–response relationship (Bodkin et al. 2003), suggesting that biological or chemical factors other than PAH exposure were affecting hepatic EROD activity (Neff, personal communication). Because the doses were high, it is also likely that even the lowest dose caused maximal induction of the CYP system (Oris, personal communication).

The results of two other studies bear on the suggestion that Harlequin Duck productivity could have been impaired by exposure to hydrocarbons. First, Stubblefield et al. (1995a, 1995b) conducted laboratory tests in which Mallards (*Anas platyrhynchos*) were fed a diet containing naturally weathered *Exxon Valdez* crude oil to determine the dosages that might affect body condition, eggshell thinning, and the like. These controlled studies established a dosage below which no effects could be detected. Second, Boehm et al. (1996) determined the PAH dosages to which Harlequin Ducks might be exposed if the birds consumed a diet with a high proportion of mussels (assumed to be 30%, shells excluded; i.e., 2–3 times known consumption rates) exclusively from oiled mussel beds. Based on this latter work, the estimated dosages they would receive 4 years after the spill were one to three orders of magnitude below the doses that produced sublethal effects in Stubblefield's Mallards. Mallards may not be a perfect surrogate for Harlequin Ducks (see Bodkin et al. 2002a; Peterson 2001), and temperature-stressed birds in nature might respond differently to a given dosage of oil than would individuals maintained under benign laboratory conditions (Rodway et al. 2003b; Esler et al. 2002; Mittelhauser 2000; Goudie and Ankney 1986). Mallards are considered the standard physiological surrogate for ducks, however (Stubblefield, personal communication), and it is unlikely that physiological differences between the species or the effects of thermal stress would override such a large difference between estimated (and almost certainly overestimated) exposure and measured sublethal dosage levels. Studies of fish have also noted the lack of clear evidence linking elevated levels of CYP1A to higher-order biological effects such as toxicity, lesions, or reproductive failure (Lee and Anderson 2005; Jewett et al. 2003).

So we are left with the observation that Harlequin Duck reproduction apparently was lower in western PWS than in eastern PWS during the 1990s, but Patten's hydrocarbon hypothesis lacks evidence on nearly every count (see Wheelwright 1994). In addition, the CYP1A evidence cannot be causally linked with either *Exxon Valdez* oil or contamination-impaired reproductive performance. Instead, the observed differences in reproduction may be related to inherent differences in habitats between the two regions of PWS.

To address the possible effects of habitat differences on Harlequin productivity, we conducted a GIS-based analysis of the availability of potential nesting habitat in the eastern and western portions of PWS (as defined by Patten and others;

Fig. 4.1a). Guided by Crowley's (1994) studies, we used stream density (streams/km of shoreline) as an indicator of overall stream availability and stream length as an index of stream discharge volume.

The differences in stream characteristics between the two areas are dramatic. We tallied 216 streams in western PWS and 207 in the eastern region, for densities of 0.14 streams/km of shoreline in western PWS (1,689 km of shoreline) and 0.21 streams/km in the east (1,035 km). Streams in the west averaged 1.9 km long, whereas those in the east averaged 5.8 km. Thus, stream density is not only ~50% greater in eastern than in western PWS, but the average length of those streams is ~300% that of streams in western PWS. Patten (1994) also noted that potential breeding streams in western PWS differed from those in the east, particularly in having lower discharge rates and being shallower and considerably shorter. Similarly, Crowley (1994) indicated that Harlequins in eastern PWS selected the largest streams available for nesting, which were streams that also were large enough for salmon to use for spawning; he found that stream discharge volume was the most important habitat factor distinguishing streams that were used from those that were not used by breeding Harlequins. Dzinbal's (1982) work also suggested that streams in PWS as a whole were lower in invertebrate productivity than were inland nesting streams. All of this led Rosenberg and Petrula (1998) to conclude that there is little suitable Harlequin breeding habitat in PWS as a whole but that, of that which is available, more occurs in eastern PWS. The Trustees (1999) stated that "the breeding habitat in the western sound is very limited compared to what is available in the eastern sound." Based simply on these differences in nesting habitat alone, one should not be surprised that reproduction and population trends differ between areas.

4.4.2.2   Survival

Quite apart from the effects of direct mortality immediately after the oil spill, the abundance and dynamics of populations could also be affected through changes in individual survival over a longer time period. Survivorship is difficult to document, but Esler et al. (2000a) conducted a 3-year study (1995–1997) in which they used radiotelemetry to assess the survival of adult female Harlequins over winter (October–March). As with several other studies, their design compared the spill area as a whole with an unoiled reference site (Montague Island). Over all years, cumulative estimated winter survival was 78.0% (SE=3.3%) in oiled areas and 83.7% (SE=2.9%) in the unoiled reference site. The difference in survival rates is not statistically significant (Parker, personal communication). Nonetheless, these survivorship estimates were applied to a Harlequin Duck population-dynamics model (Robertson 1997), which suggested that the difference in survival would be sufficient to produce a declining population trend in the oiled area that contrasted with a fairly stable population in the unoiled area. These predictions coincided with the population-trend patterns reported by Rosenberg and Petrula (1998), leading Esler et al. (2002) to conclude that the poor survival of adult females in oiled areas demonstrated continued spill effects and described "the demographic mechanisms that would lead to persistent population declines." These predictions also led Rosenberg and Petrula (1998) and Rosenberg et al. (2005) to conclude that the difference in sex ratios between oiled and unoiled parts of PWS was caused by this differential female survival. The comparison of population trends between oiled and unoiled parts of PWS, however, indicates that there is not a significantly declining population trend in the oiled area during either season (Fig. 4.4; also see McKnight et al. 2006), so the results of the population modeling of Esler et al. (2002, 2000a) may not apply.

Making this linkage between oiling history and survival requires one to assume that the unoiled reference site and the oiled capture sites were representative of the larger unoiled and oiled areas and that the survival patterns of overwintering birds carried over into the summer (i.e., they were not disrupted by differential movements of residents or nonresident breeders into oiled or unoiled areas). Because the estimates of female recruitment that Rodway et al. (2003b) derived in a study of wintering Harlequin Ducks in coastal British Columbia in 1999–2000 also suggested a declining population, Rodway et al. cautioned that "it is necessary to incorporate emigration in estimates of adult survival before demographic trends can be confidently inferred."

The conclusions of Esler et al. (2002, 2000a) about reduced survival rates in the oiled area also rest on the assumption that there are no habitat or environmental differences between the areas that would contribute to differences in survival. To evaluate this assumption, Esler et al. compared the survival of the Montague Island birds with a subset of the oiled-area birds that were captured at Green Island, reasoning that Green Island is closer to Montague Island than are the other oiled sites and, therefore, should be more similar environmentally. The survival estimate for Green Island birds (76.8%, SE=5.7%), however, more closely approximated that for all oiled areas combined than that for Montague Island. This result is not too surprising because the Green Island birds constituted 49% of the entire sample from the oiled area.

Esler et al. (2002, 2000a) speculated that the reduced winter survival of Harlequin females in the spill area was due to continuing contamination from residual oil, citing the CYP1A studies as evidence. Although the survivorship of wintering female Harlequins initially documented by Esler et al. (2002, 2000a) did differ (albeit not significantly) in different parts of PWS in a way that coincides with oiling history, it also coincided with habitat differences between the two areas. More recently, Rosenberg (cited in National Wildlife Federation 2003), Esler et al. (2003), Esler and Iverson (2010), and Bodkin et al. (2003) reported that the extent of hydrocarbon exposure and female survival rates from 2000 to 2003 were similar between oiled and unoiled areas, and Rosenberg and Petrula (2005) suggested that the recruitment rates they observed during 1997–2005 were indicative of a stable population in PWS. Overall, then, the evidence to establish a direct cause–effect linkage between oil exposure and reduced survival, much less between reduced survival and the population trends recorded by some observers (but not others), remains equivocal.

Although the hypothesized pathways and linkages that underpin the current claims of ongoing injury to Harlequin Ducks are not clearly articulated in the Trustees' most recent update on injured resources (Trustees 2009), the basic proposition appears to be that oil residues in intertidal or subtidal sediments become bioavailable through bioturbation by sea otters digging pits during their foraging

(Short et al. 2006), that Harlequins are exposed to and ingest toxic fractions of residual oil as evidenced by CYP1A studies (Esler et al. 2010), and that overwinter survival of females is reduced in the oiled regions of PWS due to this exposure (Esler et al. 2002). In the context of the exposure pathways presented in Fig. 4.2, these current claims are depicted in Fig. 4.7. Our assessment of this hypothesized pathway of exposure and injury can be summarized as follows. First, the bioturbation pathway for releasing significant quantities of bioavailable oil residues is flawed because the residual oil is buried beneath under ≥5–10 cm of clean sediment overlain by boulder and cobble veneers in the middle and upper intertidal zones, where sea otters generally do not forage (Harwell et al. in press). In addition, there are very few locations (all in Northwest Bay on Knight Island) where some residual oil is located in the lower intertidal. Bioturbation can occur only in finer grained sediments of the lower tide zone and nearshore subtidal. Second, the CYP1A results are compromised by significant environmental differences between oiled and reference areas, laboratory inconsistencies, and lack of statistical significance (Esler 2008). The form and location of residual oil in PWS and the induction of CYP1A by a variety of hydrocarbons make interpretations that link the CYP1A results directly to exposure of Harlequins to residual *Exxon Valdez* oil unrealistic. Third, the negative population trends of Harlequins in the oiled areas reported by McKnight et al. (2006) and Esler et al. (2002) are no longer detectable (this paper; Rosenberg 2008).

4.4.3 Potential Effects on Harlequin Duck Habitat Occupancy and Use

Variations in habitat appear to be important contributors to regional differences in Harlequin Duck densities and reproduction within PWS. Broadly defined, the intertidal areas and shorelines where oil was deposited by the *Exxon Valdez* spill are the primary habitat of Harlequin Ducks in PWS (except when they are actually nesting). Consequently, these habitat effects might also translate into effects on the distribution or local abundance of Harlequins. Because most of the oil had been removed from the shorelines within a year of the spill, either through the cleanup activities or by natural weathering (Neff et al. 1995), one might expect rapid reoccupancy of previously oiled habitats.

Evaluation of the effects of an oil spill on habitat occupancy and use therefore requires an explicit consideration of habitat factors as well as of shoreline oiling. Few of the studies conducted in PWS included such a consideration. Patten (1994) (Patten et al. 1995) partitioned his observations of Harlequins in western PWS among four general habitat categories, which he further subdivided among five categories of shoreline oiling intensity. During the mid-summer molting period, nearly half of his observations of Harlequins were on offshore rocks (17% of them in one large group at a single location on Channel Island), while another quarter were recorded in sheltered bays and lagoons. Few individuals were seen at the mouths of potential breeding streams. Over all habitats, 10% of the birds were seen in locations that had been heavily oiled, another 10% were in moderately oiled sites, and 37% (including the large group on Channel Island) were recorded in unoiled sites. Excluding the Channel Island group, densities were greatest near moderately and heavily oiled mussel beds and on heavily oiled offshore rocks. If Harlequins were avoiding oiled habitats at the time of these surveys, it was not readily apparent (recall, however, that data from postspill years 3–5 were combined in these analyses).

Recognizing that habitat variation among sites might confound the determination of oiling effects in BACI analyses, Irons et al. (2000) addressed the issue in two ways. First, they conducted a cluster analysis based on similarity among transects in bird-community composition and excluded from the analysis transects that were outliers in the 1984–1985 data set. This analysis was based on a single sampling of the bird community occurring in each transect; given local variation in community composition, however, there is no assurance that the transects that clustered together in 1984–1985 (and which were therefore retained for the post-spill analysis) would have clustered together in later years, independent of any disruption of the habitat. Second, Irons et al. recorded shoreline type (four categories) for each transect and used a Chi-squared analysis to determine whether the frequencies (but not amounts) of these types differed significantly between the sets of oiled and unoiled bays. Oiled transects occurred more frequently than did unoiled transects on exposed rocky shores and exposed wave-cut platforms in bedrock and less frequently on gravel beaches and sheltered rocky shores, although the differences were not significant. For a nearshore-foraging species such as the Harlequin Duck, however, these are coarse categorizations of habitat at best. Moreover, those habitats that occurred more frequently in the oiled transects are generally less suitable for Harlequins, so the differences may have been important to the birds, even if they were not statistically significant.

In another study, Esler et al. (2000b) evaluated habitat correlates of Harlequin Duck wintering densities in relation to oiling by comparing two heavily oiled locations on Knight Island with unoiled reference locations on Montague Island. Not surprisingly, their analysis indicated substantial differences in intertidal slope and coverage of rocky substrate between the areas. Wintering Harlequin densities were significantly related to several habitat variables (the presence of an offshore reef within 500 m, a stream within 200 m, and wind and wave exposure). After accounting for these relationships, however, a clear effect of location (i.e., oiling history) remained, suggesting that "population recovery from the oil spill was not complete, due either to lack of recovery from initial oil spill effects or continuing deleterious effects."

In our studies (Wiens et al. 2004; Wiens 1996; Day et al. 1995, 1997), we measured 26 features of the physical and biological habitat in ten study bays that experienced different levels of oiling (i.e., oiling was evaluated as a quantitative, rather than a categorical, variable). Overall, 11 habitat-bird density models were developed for Harlequin Ducks (one for each survey conducted during 1989–1991). Shoreline-substrate variables entered significantly into seven of these models (pebble-gravel = 4, boulder/cobble = 3, bedrock = 0).

Day et al. then used partial correlation to factor out the effects on bird densities of variation in habitat characteristics among bays. The residual values of densities were then correlated with a measure of oiling intensity to determine whether there was a significant relationship between the occupancy of sites (density) and oiling level, after accounting for habitat differences among sites. These analyses indicated that Harlequin Ducks exhibited significant negative relationships with oiling level (i.e., an oiling impact), over and above habitat-related differences, in summer 1989 and 1990. These effects disappeared by summer 1991, suggesting that recovery in habitat use was occurring. No significant relationships were detected in analyses for spring, fall, or winter data, indicating that impacts were limited to the breeding/summering population and that the habitat was occupied at other times of year without regard to the level of oiling. Parallel analyses that included surveys in mid-summer 1996, 1998, and 2001 (Wiens et al. 2004) indicated that, when only oiling level was included in the analysis, significant negative impacts of oiling on Harlequin densities occurred in mid-summer 1989–1991 (and in the 1984 prespill data as well) but not thereafter (Fig. 4.9). When habitat variations among the bays were considered, however, all significant relationships with oiling disappeared. The negative relationships with oiling level evident in most years apparently reflected subtle but systematic habitat differences among the study bays, rather than their oiling history per se.

Several assumptions underlie this approach to gauging habitat occupancy and interpreting the results. First, it assumes that the reduced use of oiled habitats reflects either direct mortality of birds in oiled areas (without immediate immigration of replacement individuals) or active avoidance of oiled areas (i.e., emigration). Second, it assumes that there is a quantitative relationship between the level of oiling of a bay and the suitability of the habitat in that bay to the birds (although this relationship need not be linear; thus, we included non-linear terms in the analyses). Third, it assumes that any site fidelity by individuals that used an area before the spill will be overridden by the dramatic changes in habitat conditions. Fourth, it assumes that subsequent reoccupation of oiled sites that previously had exhibited reduced occupancy (i.e., a spill impact) indicates that the suitability of the habitat (as viewed by the birds) has improved. Finally, it assumes that entire bays represent an appropriate scale (i.e., intermediate scale) at which to gauge spill effects on this species. In fact, oiling levels varied along shorelines within bays (Day et al. 1997; Neff et al. 1995). Esler et al. (2000b) based their study on surveys of shoreline segments of 200 m rather than entire bays, reasoning that the tight site fidelity of wintering Harlequins justified analysis on such a fine scale. They attributed the differences between the findings of their study and those of Day et al. to the differences in scale, even though the scale they used was an order of magnitude smaller than that used by the birds (Iverson et al. 2004). On the other hand, the use of short segments of shoreline as sample units by Esler et al. also increased their sample sizes by an order of magnitude over those of Day et al., enhancing their ability to detect subtle effects.

Most of the assumptions of the Day et al. approach (as well as that of Esler et al.) relate to the interplay between site fidelity and habitat selection. Many Harlequin individuals do show strong fidelity to molting and wintering locations (Iverson and Esler 2006; Iverson et al. 2004), and there is evidence of fidelity to breeding sites as well.

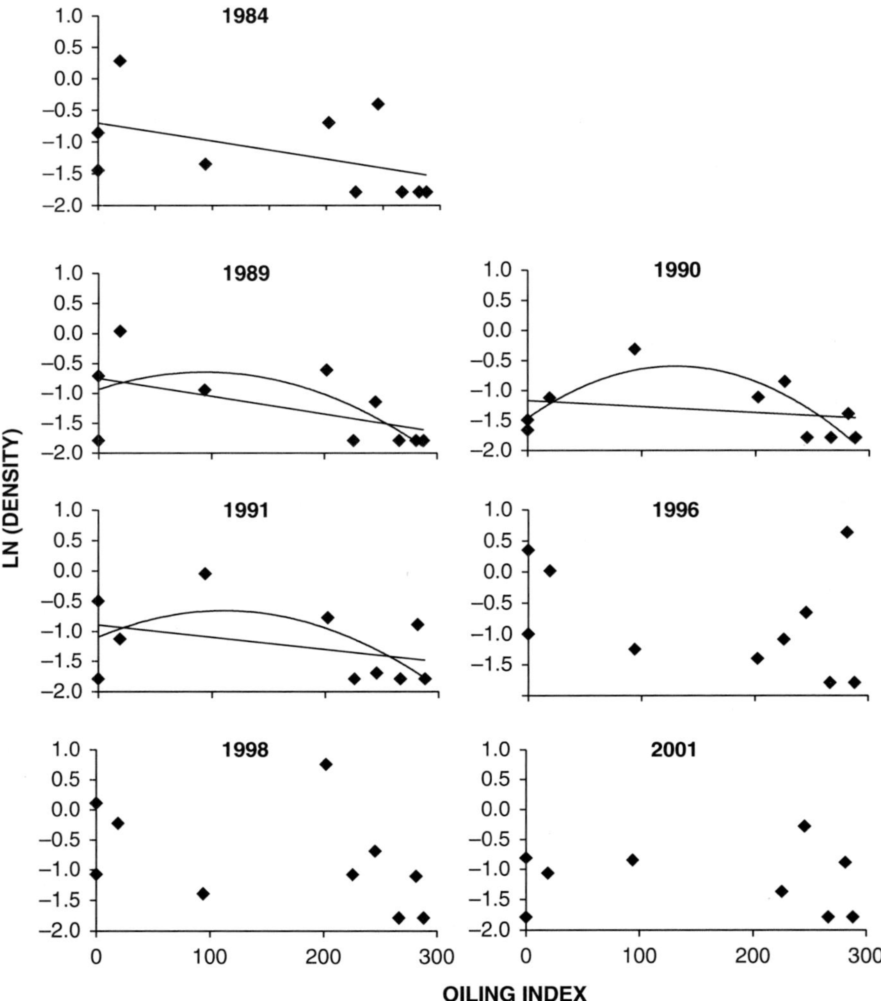

Fig. 4.9 Regressions of Harlequin Duck mid-summer density against an index of oiling (0=no oil, 400=extreme oiling) for ten study bays in Prince William Sound. Significant linear or quadratic relationships are shown; otherwise, neither relationship was significant (α=0.20)

The extent of site fidelity differs among sexes and age classes, however (Robertson et al. 2000). Because it is easier to record limited movement than broad-scale dispersal, the interpretation of such information is not easy. Even less is known about habitat selection by marine-oriented birds. In particular, what cues are used in determining whether a location is suitable or not, and at what level does oiling of the beaches, or human disturbance associated with cleanup activities, render a habitat unsuitable? What criteria do birds use to assess whether or not a previously oiled habitat is now "recovered?" In the absence of clear answers to such questions, we must resort to using habitat occupancy by the birds as *their* bioassay of overall habitat condition.

4.5 Discussion and Conclusions

4.5.1 Assessing the Effects of the Exxon Valdez Oil Spill on Harlequin Ducks

Confronted with the task of evaluating the consequences of an environmental disruption such as an oil spill, one must ask two questions. The first is, "are there ecological effects?" In this case, do we see ecological changes that are consistent with the likely impacts of the *Exxon Valdez* oil spill?

There is no doubt that several hundred Harlequin Ducks were killed directly by the spill. Moreover, there are indications that the overwinter survival of females was lower in oiled than in unoiled parts of western PWS, at least initially. Whether these consequences have translated into differences in Harlequin abundance or population trends that are causally linked to the oil spill is more problematic. Harlequin abundances increased over PWS as a whole from the 1970s to the 1990s. Some analyses suggest that the rate of increase was greater in unoiled than oiled areas, leading to an increasing divergence in relative abundances that could be interpreted as a continuing oil-spill effect. Other analyses, however, indicate that the rate of increase was greater in oiled than in unoiled areas, suggesting population recovery in the oiled areas. Yet other analyses suggested that the increase in the unoiled area was driven more by increases in the glaciated fjords than in locations more similar to the oiled areas, and others suggested no statistical differences in trends between the two areas. It does appear that densities were generally lower in oiled areas than in unoiled areas, both over PWS as a whole and within the oil-spill area. However, the same relationship existed prior to the spill, so it is hard to conclude that the differences between areas were a direct consequence of the oil spill. Considerations of habitat occupancy in relation to oiling suggested that Harlequins did not conspicuously avoid heavily or moderately oiled areas years after the spill, although there was evidence of a statistically significant negative relationship with the intensity of oiling in study bays during the summers of 1989–1991 and a lower abundance of wintering birds in oiled than in unoiled areas. The negative relationship with (subsequent) bay oiling in summer and its relationship to habitat differences among the bays was also seen in 1984, 5 years before the spill occurred, so the lower wintering abundance in oiled areas would appear more likely to be related to habitat suitability than to oiling effects.

Similar uncertainties and inconsistencies appear when information on reproduction is evaluated. Counts of broods suggest that reproductive output was lower after than before the spill in some areas, although there are doubts that the prespill surveys accurately identified Harlequin broods in some locations. Postspill surveys indicate that brood counts were lower in western than in eastern PWS, which is consistent with an oiling effect. It is clear, however, that survey efforts were neither equally nor evenly distributed over the two regions, and some scientists have questioned whether brood counts really do provide an adequate index of reproductive success. It also is apparent that there is generally less reproductive activity in western PWS than in the east because of habitat differences.

The second question we must ask to evaluate the consequences of an oil spill is, "are the effects we see due to oil, or are they due to something else?" Because of the expectation that an oil spill will have strong negative effects, there is a tendency to presume at the outset that any negative differences between oiled and unoiled areas result from the spill. One does not want to discount the possible impacts of an oil spill too easily; this is why avoidance of Type II errors is favored over the customary approach of minimizing Type I errors in statistical tests (Wiens and Parker 1995). Still, observing effects that are consistent with what we expect from an oiling impact does not necessarily mean that the effects were actually caused by the spill. Inconsistencies in apparent spill effects may arise because there are multiple pathways by which an oil spill can affect environments and populations (Fig. 4.2) and because not all mechanisms are likely to operate in the same ways or with the same magnitudes in all spill-affected areas. To be confident that there is a causal link between an oil spill and an apparently consistent effect, the mechanisms linking the two must be specified, they must be biologically plausible, and they must be supported by evidence. These criteria are not met in the mechanistic pathway linking oiling to the apparent reproductive failure or survival of Harlequins through hydrocarbon effects (Fig. 4.7).

To deal with apparent spill-related effects objectively, it is important to consider hypotheses that may link the observed effects to causes other than oiling. For many of the possible spill-related effects observed for Harlequin Ducks following the *Exxon Valdez* spill, systematic differences in habitat conditions between oiled and unoiled areas may explain the observed patterns as well as does the presence of oil from the spill. Common sense suggests that both oiling and underlying habitat characteristics may be involved, but, because the two factors may covary, it is difficult to separate their effects and to designate either as the unequivocal cause of differences between areas in abundance, population dynamics, reproduction, or habitat occupancy (Wiens and Parker 1995).

It is also important to consider the effects of the scales of time and space on observations and interpretations. For example, unless one accepts the assumption that the affected system is in a long-term equilibrium, comparisons with prespill "baseline" data may be problematic, and such comparisons become increasingly unreliable as the time gap between the pre and postspill data widens (Wiens et al. 2001b; Paine et al. 1996; Wiens 1996; Boersma et al. 1995). Assuming equilibrium in any ecological system is questionable, but this may be especially true for high-latitude marine ecosystems. For example, PWS and the Gulf of Alaska underwent several broad-scale environmental shifts during the decades preceding the spill: the 9.2-magnitude Alaska earthquake in 1964 and subsequent release of petroleum hydrocarbons from storage tanks at Valdez and other nearby communities; oceanic regime shifts in the late 1970s and 1989; the 1976–1977, 1982–1983, and 1997–1998 *El Niño* events; and an exceptionally cold winter just before the spill (including the coldest temperatures ever recorded in Valdez). These events undoubtedly affected marine birds and their habitats (Holloway 1996; Paine et al. 1996; Boersma et al. 1995). In addition, this region has experienced long-term oceanographic changes (Hare and Mantua 2000; Francis et al. 1998; Hayward 1997) and fundamental changes in the abundances of fishes on which seabirds feed have occurred as well (Peterson and Schwing 2003; Agler et al. 1999; Piatt and Anderson 1996), suggesting that long-term changes in the invertebrate communities on which Harlequin Ducks feed probably also have occurred. The PWS ecosystem is clearly dynamic (for a 2000-year perspective on environmental changes in this region, see Finney et al. 2002).

So, when all is said and done, what *can* be concluded about the effects of the *Exxon Valdez* oil spill on Harlequin Ducks? Citing concerns about possible oil exposure, reduced survival, and a possible ongoing decline in numbers of Harlequins in western PWS, the Trustees (2009) concluded that the available observations suggest that Harlequin Ducks are recovering, but have not fully recovered from the effects of the oil spill. On the other hand, Day et al. (1997), Murphy et al. (1997), and Wiens et al. (2004) concluded that Harlequin Ducks either showed no evidence of spill impacts on patterns of habitat occupancy or had recovered from any impacts on habitat use by 1991. Although Irons et al. (1999) stated that no effects on Harlequin Duck populations were evident after 1991, Irons et al. (2000) nonetheless concluded that Harlequin Ducks exhibited "strong evidence of negative oil spill effects a few years after the spill and may be recovering" (but see Wiens et al. 2001b). Similarly, McKnight et al. (2006) concluded that population data from 2005 "did not show any evidence of a recovering population." In a broader consideration of the biological effects of the *Exxon Valdez* spill, Paine et al. (1996) concluded that few of the studies that assessed changes in population density or demography in PWS "gave incontrovertible evidence of an oil effect."

4.5.2 What Broader Lessons Can Be Learned from this Case Study?

The example of Harlequin Ducks and the *Exxon Valdez* oil spill would seem to provide an ideal situation in which to reach firm and unambiguous conclusions about the effects of an environmental accident – a conspicuous and charismatic species closely tied to the shoreline habitat that was so clearly contaminated by the largest tanker oil spill in North America and subjected to a massive research effort. Yet reaching such conclusions seems elusive, not just for Harlequin Ducks, but for a host of other organisms as well (e.g., shoreline invertebrates: Gilfillan et al. 2002; Peterson et al. 2002; Skalski et al. 2001; Peterson et al. 2001; Gilfillan et al. 1995a, 1995b; Page et al. 1995; subtidal communities: Dean et al. 2000, 2001; *Fucus:* Driskell et al. 2001; other bird species: Golet et al. 2002; Boersma and Clark 2001; Wiens et al. 2001b, 2004; Lance et al. 2001; Irons et al. 2000; Day et al. 1997; mussels: Carls et al. 2001, 2004; Boehm et al. 1996; 2004; salmon: Rice et al. 2001; Brannon and Maki 1996; herring: Carls et al. 2002; Pearson et al. 1999; Hose et al. 1996; Kocan et al. 1996; sea otters: Dean et al. 2002; Bodkin et al. 2002b; Garshelis and Johnson 2001; Monson et al. 2000; Lance et al. 1999; Garshelis 1997; Johnson and Garshelis 1995; Burn 1994; Garrott et al. 1993; river otters: Bowyer et al. 2003; and harbor seals: ver Hoef and Frost 2003; Hoover-Miller et al. 2001; see also Harwell and Gentile 2006; Peterson 2001; Paine et al. 1996). Why is this so?

The simple answer is that, in a variable, scale-dependent ecosystem such as PWS (and, ultimately, all ecosystems), studies that address different questions with differing preconceptions and assumptions and that are designed and analyzed in different ways can yield different conclusions. Further, the longer such studies are conducted, the more difficult it becomes to determine which conclusion is "right." These are issues that plague attempts to assess the ecological impacts of any large or chronic disruptions of the environment, whether they are due to natural events such as cyclones, floods, or volcanic eruptions or to anthropogenic disturbances such as grazing, forest clearing, or eutrophication of streams and lakes. We see several lessons that emerge from the Harlequin Duck case study that may be relevant to these broader concerns.

4.5.2.1 The Importance of the Question Asked

There are two ways (at least) to pose a question about the impact of some environmental disruption. One can ask either "were there effects?" or "how bad was it?" The first question includes the possibility that there may have been no effects, whereas the second not only assumes that there were effects, but that those effects were negative. The difference between the questions is important, for it determines how a study is designed and what sorts of statistics (if any) are used to analyze the results. The nature of the question therefore determines and constrains what sorts of answers can be obtained. Because many environmental disruptions (especially dramatic ones such as oil spills) kindle emotional responses, the preconceptions about effects that predicate the second question can produce responses that are as much advocacy as they are science (Wiens 1997, 1996).

4.5.2.2 The Importance of the Study Design

Because an accident such as an oil spill represents such an obvious alteration of the environment, it is tempting to think of it as an "experiment" of sorts and design a study accordingly. The affected area or some subset of it then becomes the "treatment," which is compared with an unaffected "control" (or, more appropriately, "reference") area. The parallel with a true experiment, however, is tenuous at best. Not only is the treatment unreplicated (which is fortunate), but rarely does it occur randomly in relation to preexisting environmental conditions. Typically, environmental factors other than the treatment itself are not considered or measured. One then must assume that such factors are unimportant, are overwhelmed by the effects of the treatment, or vary randomly with respect to the treatment. As we have indicated above, none of these assumptions is likely to be valid. If a study design uses reference areas that are incorrectly classified with respect to the "treatment" or without considering the effects of potentially confounding variables, investigators may end up not testing what they thought they were testing, and their conclusions will be suspect.

It is also rare that an environmental perturbation acts in an all-or-none fashion. Rather, the magnitude of the environmental disturbance and its effects on the environment will vary with the level of disturbance. Because the effects of these variations in intensity on habitats and organisms may be "dose-dependent" or nonlinear (i.e., threshold), it may be more appropriate to include evaluation of such variations explicitly as part of the study design.

Overall, a variety of designs and approaches is possible (Parker and Wiens 2005; McDonald et al. 2000; Longpré et al. 1997; Schmitt and Osenberg 1996; Wiens and Parker 1995). It is not that one of these study designs is "best" under all conditions, although some are clearly better than others. Logistical constraints and the lack of replication and randomization often preclude some of the more powerful designs, yet the study design dictates how data will be collected and how they will be compared and evaluated. Therefore, different study designs can yield different conclusions. The study design needs to be carefully matched to the question(s) being asked and the analyses to be done, and its inherent biases and limitations need to be acknowledged and examined at the outset.

4.5.2.3 The Importance of Variance and Scale

Designing studies to assess impacts and recovery from environmental perturbations is difficult in part because the "treatment" is superimposed on an environment that varies in both space and time. Unfortunately, in such cases, treatment and control areas often are different – they contain different complexes of environmental factors and undergo different dynamics, quite apart from the perturbation that is the focus of study. Moreover, as more time passes following a 1-time, "pulse" perturbation, more things happen, many of which are unrelated to the perturbation itself. Both of these factors increase the "noise" in a system over time. This temporal variation makes it difficult to establish unambiguously what factors or processes are responsible for the observed patterns and to distinguish "signal" from "noise" (which may actually reflect many ecologically important processes and interactions).

As in all areas of ecology, determining causal relationships is also complicated by scale. Whether one assesses oil-spill impacts over tens or hundreds of meters of shoreline, over bays or other similarly large units, or over the spill area as a whole affects the ways in which data are aggregated and averaged. If the sample units that are used in a study are so large that they include a mix of impacted and unaffected areas, for example, the within-sample heterogeneity may confound the results. Details that are important at one scale may fade into the background at other scales. If investigators conduct their studies at different scales, they should not be surprised if their conclusions do not match.

4.5.2.4   The Importance of Considering Multiple Hypotheses

Testing of multiple hypotheses has been an intermittent part of the scientist's toolbag since the mid-nineteenth century (Chamberlain 1890; see also Burnham and Anderson 1998; Platt 1964). Despite this capability, ecologists often still seek single-factor explanations for observed patterns. This preference for single hypotheses is especially true of large environmental disruptions in which the causal agent is obvious and seemingly overpowering. Because various other factors were acting on the ecological system of interest to produce patterns before the perturbation, however, it is naïve to presume that these factors or causal pathways are no longer important simply because a perturbation has occurred. For example, sea otters were once overharvested in PWS; then repopulation occurred well before the *Exxon Valdez* spill. Because different areas of the Sound were repopulated at different times, population trends in these areas differed. The oil spill was superimposed on these local dynamics and on a backdrop of earthquake history and prey population dynamics (Garshelis and Johnson 2001; Johnson, personal communication).

Regardless of the details of study design, differences in habitat, prey availability, predation risk, background levels of human disturbance, land- (or water-) use practices, and a host of other factors may occur among the areas being compared. Some of these factors may have clear logical relationships to the response variables being measured. To exclude such factors from consideration increases the risk of concluding that the effects are due to the perturbation alone, when the explanatory power of other possible causes may be as great or greater. Preconceptions about perturbation effects may foster a false sense of security in tests of unitary hypotheses. The real question is whether "other factors" should be considered as contributing causal variables in analyses or only as potential confounding variables. Development of causal explanations of the effects of environmental disruptions should include consideration of multiple factors, perhaps by using a formalized model-selection procedure (e.g., Burnham and Anderson 1998).

4.5.2.5   The Importance of Defining Terms Operationally

The primary concern in evaluating the environmental impacts of an event usually is in determining whether there were impacts and, if so, whether and when recovery occurs. This assessment requires clear, biologically sound, and operational definitions of "impact" and "recovery." Broadly speaking, "impact" is a departure from normal conditions and "recovery" is a return to those conditions. But what is regarded as "normal" may ultimately boil down to one's philosophy of Nature. Adherents to a "balance of nature" philosophy may hold that the system is in a steady-state equilibrium (Wiens 1977), thus specifying a stable reference condition. By this criterion, any departure from the conditions that existed before the perturbation can be taken as an impact, and recovery occurs only when the system returns to the preperturbation reference conditions. In the case of the *Exxon Valdez* oil spill, for example, the Trustees (1994) stated that "full ecological recovery will have been achieved when populations of flora and fauna are again present at former or prespill abundances, healthy and productive and there is a full complement of age classes at the level that would have been present had the spill not occurred." By this definition, Harlequin Ducks will have recovered "when breeding- and nonbreeding-season densities return to prespill levels" (Trustees 1999), "when hydrocarbon exposure is similar between oiled and unoiled areas; when numbers are stable or increasing; and when demographic attributes are similar and densities return to prespill levels" (Trustees 2002b), or "when breeding- and nonbreeding-season demographics return to prespill levels and when biochemical indicators of hydrocarbon exposure of harlequins in oiled areas of Prince William Sound are similar to those in harlequins in unoiled areas" (Trustees 2006). Esler et al. (2002) linked full population recovery of Harlequin Ducks to a "return to prespill numbers following cessation of residual oil spill effects."

This view is clearly founded on an equilibrium perception – i.e., in the absence of the spill, populations would have been as they were before the spill – and it follows directly from legislation governing the assessment of damages from oil spills or other pollution of aquatic ecosystems [e.g., CERCLA (42USC9601) and OPA (33USC2701)]. All ecological systems are variable in time and space, however, and any "equilibrium" is often coarse at best. Under these conditions, it is not possible to specify a single value as a baseline against which to judge impact or recovery (see Parker and Wiens 2005; Kingston 2002). Moreover, like almost any ecological system, PWS has been subjected to a long history of human exploitation and impacts, so the environment prior to the spill was scarcely "pristine" (Wooley 2002). What, then, is the measure of "impact" or "recovery" in a variable and previously impacted system? A departure from and return to "average" or "normal" conditions? But over what time period or area is the "average" to be calculated, and how much departure from "normal" represents an impact?

4.5.2.6 The Importance of Statistics and Natural History

Operationally, impact and recovery in a variable system can be defined using statistical criteria (e.g., Wiens 1995): a significant departure from baseline or reference conditions indicates an impact, and the disappearance of a previously significant effect represents recovery. [This approach has been criticized by Irons et al. (2001), who argued that it sets a lower standard for documenting recovery than for impact. We fail to see the logic of this argument, as the statistical tests we used established a "significance window" (defined by $\alpha = 0.20$) that is just as difficult to enter (i.e., indicating impact) as it is to leave (i.e., indicating recovery).] This statistical approach is operational, but it is also sensitive to the constraints of any statistical test. The detection of a statistically significant difference between populations (e.g., a disturbed and an undisturbed area) is related to the intrinsic variability of the system (the among-sample variance), the sample size, and the amount of difference (i.e., effect size). Thus, the greater the variability of the system in time or space, the greater the difference required to attain a given level of statistical significance, but the greater the sample size or effect size, the less difference required.

Ecologists often attempt to maximize sample sizes within logistical constraints to detect small differences statistically. To some degree, this approach reflects a continuing adherence to equilibrium thinking – one compensates for variance by increasing sample size so that even small departures from "normal" conditions can be detected. Ultimately, of course, the objective is to detect changes or differences that are *biologically* significant, which may or may not be recognized statistically (Hatch 2003). This is where thinking about effect size comes in. Statistical power relates to the capacity to detect an effect of a certain size with a given variance and sample size, and power is now routinely calculated in assessments of the consequences of environmental perturbations (e.g., Irons et al. 2000; Murphy et al. 1997). It is more important, however, to think about *biological power*: what effect size is likely to have repercussions on the biological functioning of the system of interest? For example, does a change in abundance or reproductive output of 1 or 10% represent a biologically significant change, regardless of whether it can be judged to be statistically significant? The answer depends on both the natural variability of the system and the natural history of the organisms of interest. Unfortunately, in most situations, neither of these issues is known in quantitative detail, so judgments of impacts and recovery often contain an element of subjectivity. This subjectivity is why statistical approaches, especially those based on equilibrium criteria, are particularly attractive. To ignore natural-history insights in the rush to be quantitative and objective, however, runs the risk of reaching statistically rigorous but biologically irrelevant conclusions.

4.5.2.7 The Importance of Science in Assessing Environmental Impacts

When confronted with an environmental accident or disruption, it is important to bring the tools, methods, and honesty of science to bear on the issue of documenting effects and determining causes. The alternative is often a subjective evaluation that is colored by emotions and preconceptions and that, at worst, can be distorted by agendas and advocacy (e.g., Ott 2005). In such situations, the problem is that many of the usual tools of science – experiments, balanced study designs, straightforward statistical analyses, empirically based models, and the like – are difficult to use. Consequently, the results of scientific studies may contain more uncertainty than one might like. At the very time that the general public and legal proceedings are demanding clear answers to questions about impacts and recovery, science seems plagued by uncertainties, ambiguities, and inconsistencies. Many of these problems are evident in the Harlequin Duck case study that we have presented here, and they may be exacerbated by ignoring the points detailed above. Science, however, is more than a set of tools to be applied to a problem. It is an *approach*, a way of gaining knowledge that is characterized by objectivity, questioning of preconceptions, embracing the unexpected, providing an honest accounting of assumptions, methods, and results, and couching interpretations and conclusions in the context of the quantity and quality of the information at hand. Attention to the points we have discussed above should do much to achieve these objectives. Even if reality imposes severe constraints on the use of traditional scientific tools, it need not constrain careful thinking.

One can also legitimately ask how much certainty is required to establish likely cause–effect relationships in such a dramatic event as an oil spill. Where should the burden of proof lie? It may be appropriate to adopt a conservative approach to judging impacts and recovery to avoid Type II statistical errors – failing to detect effects that are really there (Longpré et al. 1997; Wiens and Parker 1995; Shrader-Frechette and McCoy 1993). One should view a failure to reject a null hypothesis of no effect with careful skepticism. It may be more appropriate to approach the issue of impacts and recovery from a variety of perspectives using a variety of analyses and then employ a "weight-of-evidence" approach in drawing conclusions. The conclusions should also be limited to what the data and analyses really show, not what one thinks they should show. Drawing broad conclusions about impacts and recovery based on possible effects with equivocal support derived from searching for confirmatory evidence rather than careful sampling and analysis is contrary to the spirit of science.

There is an additional, confounding factor. Scientific uncertainty opens the door to conscious or subconscious advocacy. Environmental disruptions such as oil spills are emotional events that offend the sensibilities of people concerned about the state of the environment. Common sense dictates that such events will have profound and far-reaching effects. Given this preconception, it is easy to see in one's observations what one expects (or wants) to see, to dispense with the tedium of broad-scale sampling and rigorous statistical analysis, to ignore possible alternative hypotheses, and to reach premature conclusions based on shreds of evidence. Large environmental accidents such as the *Exxon Valdez* spill attract wide public and media attention, and, if litigation enters the picture, positions can rapidly become polarized and hardened, making it difficult to change one's conclusions. As a result, groups other than scientists may grasp onto the results they want, stripping them of their scientific uncertainty and ignoring contrary findings. Unsubstantiated speculations can easily be converted into established facts (e.g., Chadwick 1993; Pain 1993; Keeble 1991; but see Wheelwright 1994). Alternatively, public opinion can override clear evidence of recovery to cause a species to continue not being classified as recovered (e.g., the Black Oystercatcher *Haematopus bachmani*; compare Trustees 2006; with Murphy and Mabee 2000). In such circumstances, it would be wise for scientists to avoid reaching premature conclusions from limited (or no) data, to withhold judgment until more complete data have been collected and analyzed, and only then discuss the findings with the media or the public. It is likely that a clearer view of the effects of the *Exxon Valdez* oil spill on Harlequin Ducks would have emerged if the picture had not been clouded by some of the statements made soon after the spill (Table 4.1), which some individuals then felt compelled to defend, despite contrary evidence.

Acknowledgments Many individuals have contributed to this synthesis. Alan Maki, Alan Mut, Keith Parker, and a large number of ABR personnel assisted in field work and data management. Allison Zusi-Cobb prepared GIS graphics. Charles (Rick) Johnson assisted in field work and contributed insights gained from his work on sea otters. Joint field work with Dan Rosenberg in 1996 provided fresh perspectives on Harlequin Duck ecology. This work was initiated while JAW was associated with Colorado State University, and preparation of the manuscript was facilitated while JAW was a Sabbatical Fellow at the National Center for Ecological Analysis and Synthesis, a Center funded by NSF (Grant # DEB-0072909), the University of California, and UC Santa Barbara. Development of the manuscript continued while JAW was a Lead Scientist with The Nature Conservancy. We thank all of these individuals and organizations. Our research was funded by Exxon Company, USA, and by ExxonMobil Corporation; however, the conclusions are our own and do not necessarily represent those of ExxonMobil.

References

Agler BA, Kendall SJ (1997) Marine bird and sea otter abundance of Prince William Sound, Alaska: trends following the T/V *Exxon Valdez* oil spill, 1989–96, *Exxon Valdez* oil spill Restoration Final Report (Restoration Project 96159). Exxon Valdez Oil Spill Trustee Council, Anchorage

Agler BA, Seiser PE, Kendall SJ, Irons DB (1994) Marine bird and Sea Otter population abundance of Prince William Sound, Alaska: trends following the *T/V Exxon Valdez* oil spill, 1989–93, *Exxon Valdez* oil spill *Exxon Valdez* oil spill Restoration Final Report (Restoration Project 93045). Exxon Valdez Oil Spill Trustee Council, Anchorage

Agler BA, Seiser PE, Kendall SJ, Irons DB (1995) Winter marine bird and sea otter abundance of Prince William Sound, Alaska: population trends following the *T/V Exxon Valdez* oil spill from 1990–94 [sic], *Exxon Valdez* oil spill Restoration Final Report (Restoration Project 94159). Exxon Valdez Oil Spill Trustee Council, Anchorage

Agler BA, Kendall SJ, Irons DB, Klosiewski SP (1999) Declines in marine bird populations in Prince William Sound, Alaska, coincident with a climatic regime shift. Waterbirds 22:98–103

Babcock MM, Irvine GV, Harris PM, Cusick JA, Rice SD (1996) Persistence of oiling in mussel beds three and four years after the *Exxon Valdez* oil spill. In: Rice SD, Spies RB, Wolfe DA, Wright BA (eds) Proceedings of the *Exxon Valdez* Oil Spill Symposium, Symposium No. 18. American Fisheries Society, Bethesda, pp 286–297

Bence AE, Burns WA (1995) Fingerprinting hydrocarbons in the biological resources of the *Exxon Valdez* spill area. In: Wells PG, Butler JN, Hughes JS (eds) *Exxon Valdez* oil spill: fate and effects in Alaskan waters, ASTM Special Technical Publication No. 1219. American Society of Testing and Materials, Philadelphia, pp 84–140

Bence AE, Page DS, Boehm PD (2007) Advances in forensic techniques for petroleum hydrocarbons: the *Exxon Valdez* experience. In: Wang Z, Stout SA (eds) Oil spill environmental forensics. Fingerprinting and source identification. Academic, San Diego, pp 449–487

Bengtson S-A (1972) Breeding ecology of the Harlequin Duck *Histrionicus histrionicus* (L.) in Iceland. Ornis Scandinavica 3:1–19

Bengtson S-A, Ulfstrand S (1971) Food resources and breeding frequency of the Harlequin Duck *Histrionicus histrionicus* in Iceland. Oikos 22:235–239

Bodkin JL, Ballachey BE (2004) Lingering oil and sea otters: pathways of exposure and recovery status (Continuation of project 030620). GEM Proposal. Exxon Valdez Trustee Council, Anchorage

Bodkin JL, Ballachey BE, Esler D (2002a) Patterns and processes of population change in selected nearshore vertebrate predators, *Exxon Valdez* oil spill Restoration Project Annual Report (Restoration Project 01423). Exxon Valdez Oil Spill Trustee Council, Anchorage

Bodkin JL, Ballachey BE, Esler D, Dean T (2003) Patterns and processes of population change in selected nearshore vertebrate predators, *Exxon Valdez* oil spill Restoration Project Final Report (Restoration Project 030423). Exxon Valdez Oil Spill Trustee Council, Anchorage

Bodkin JL, Ballachey BE, Dean TA, Fukuyama AK, Jewett SC, McDonald L, Monson DH, O'Clair CE, VanBlaricom GR (2002b) Sea otter population status and the process of recovery from the 1989 '*Exxon Valdez*' oil spill. Mar Ecol Prog Ser 241:237–253

Boehm PD, Page DS, Gilfillan ES, Stubblefield WA, Harner EJ (1995) Shoreline ecology program for Prince William Sound, Alaska, following the *Exxon Valdez* oil spill: part 2 – chemistry and toxicology. In: Wells PG, Butler JN, Hughes JS (eds) *Exxon Valdez* oil spill: fate and effects in Alaskan waters, ASTM Special Technical Publication No. 1219. American Society of Testing and Materials, Philadelphia, pp 347–397

Boehm PD, Mankiewicz PJ, Hartung R, Neff JM, Page DS, Gilfillan ES, O'Reilly JE, Parker KR (1996) Characterization of mussel beds with residual oil and the risk to foraging wildlife 4 years after the *Exxon Valdez* oil spill. Environ Toxicol Chem 15:1289–1303

Boehm PD, Neff JM, Page DS (2003) Letter to the editor. Mar Environ Res 55:459–461

Boehm PD, Page DS, Brown JS, Neff JM, Burns WA (2004) Polycyclic aromatic hydrocarbon levels in mussels from Prince William Sound, Alaska, USA, document the return to baseline conditions. Environ Toxicol Chem 23:2916–2929

Boehm PD, Page DS, Brown JS, Neff JM, Bragg J, Atlas R (2008) Distribution and weathering of crude oil residues on shorelines 18 years after the *Exxon Valdez* spill. Environ Sci Technol 42:9210–9216

Boersma PD, Parrish JK, Kettle AB (1995) Common Murre abundance, phenology, and productivity on the Barren Islands, Alaska: the *Exxon Valdez* oil spill and long-term environmental change. In: Wells PG, Butler JN, Hughes JS (eds) *Exxon Valdez* oil spill: fate and effects in Alaskan waters, ASTM Special Technical Publication No. 1219. American Society for Testing and Materials, Philadelphia, pp 820–853.

Boersma PD, Clark JA (2001) Seabird recovery following the *Exxon Valdez* oil spill: why was murre recovery controversial? Proceedings 2001 International Oil Spill Conference. API Publication 14710. American Petroleum. Institute, Washington, DC, pp 1509–1514

Bowyer RT, Blundell GM, Ben-David M, Jewett SC, Dean TA, Duffy LK (2003) Effects of the *Exxon Valdez* oil spill on river otters: injury and recovery of a sentinel species. Wildl Monogr 153:1–53

Brannon EL, Maki AW (1996) The *Exxon Valdez* oil spill: analysis of impacts on the Prince William Sound pink salmon. Rev Fish Sci 4:289–337

Burger J (1997) Oil spills. Rutgers University, New Brunswick

Burn DM (1994) Boat-based population surveys of sea otters in Prince William Sound. In: Loughlin TR (ed) Marine mammals and the *Exxon Valdez*. Academic, San Diego, pp 61–80

Burnham KP, Anderson DR (1998) Model selection and inference: a practical information-theoretical approach. Springer, New York

Carls MG, Babcock MM, Harris PM, Irvine GV, Cusick JA, Rice SD (2001) Persistence of oiling in mussel beds after the *Exxon Valdez* oil spill. Mar Environ Res 51:167–190

Carls MG, Marty GD, Hose JE (2002) Synthesis of the toxicological impacts of the *Exxon Valdez* oil spill on Pacific herring (*Clupea pallasi*) in Prince William Sound, Alaska, U.S.A. Can J Fish Aquat Sci 59:153–172

Carls MG, Harris PM, Rice SD (2004) Restoration of oiled mussel beds in Prince William Sound, Alaska. Mar Environ Res 57:359–376

Cassirer EF, Groves CR (1994) Ecology of Harlequin Ducks *Histrionicus histrionicus* in northern Idaho, Unpublished Technical Report. Idaho Department of Fish and Game, Boise

Chadwick DH (1993) Harlequin ducks. Natl Geogr 184(5):116–132

Chamberlain TC (1890) The method of multiple working hypotheses, Reprinted in 1965. Science 148:754–759

Clark RB (1984) Impact of oil pollution on seabirds. Environ Pollut Ser A 33:1–22

Clarkson P (1992) A preliminary investigation into the status and distribution of Harlequin Ducks in Jasper National Park, Natural Resource Conservation. Jasper National Park, Jasper

Cooke F, Robertson GJ, Smith CM, Goudie RI, Boyd WS (2000) Survival, emigration, and winter population structure of Harlequin Ducks. Condor 102:137–144

Crowley DW (1994) Breeding habitat of Harlequin Ducks in Prince William Sound, Alaska. M.S. thesis, Oregon State University, Corvallis

Crowley DW (1995) Breeding ecology of Harlequin Ducks in Prince William Sound, Alaska, *Exxon Valdez* Oil Spill Final Report (Restoration Project 71). Exxon Valdez Oil Spill Trustee Council, Anchorage

Day RH, Murphy SM, Wiens JA, Hayward GC, Harner EJ, Smith LN (1995) Use of oil-affected habitats by birds after the *Exxon Valdez* oil spill. In: Wells PG, Butler JN, Hughes JS (eds) *Exxon Valdez* oil spill: fate and effects in Alaskan waters, ASTM Special Technical Publication No. 1219. American Society of Testing and Materials, Philadelphia, pp 726–761

Day RH, Murphy SM, Wiens JA, Hayward GD, Harner EJ, Smith LN (1997) Effects of the *Exxon Valdez* oil spill on habitat use by birds in Prince William Sound, Alaska. Ecol Appl 7:593–613

Dean TA, Bodkin JL, Jewett SC, Monson DH, Jung D (2000) Changes in sea urchins and kelp following a reduction in sea otter density as a result of the *Exxon Valdez* oil spill. Mar Ecol Prog Ser 199:281–291

Dean TA, Jewett SC, McIntosh PD (2001) Habitat-specific recovery of shallow subtidal communities following the *Exxon Valdez* oil spill. Ecol Appl 11:1456–1471

Dean TA, Bodkin JL, Fukuyama AK, Jewett SC, Monson DH, O'Clair CE, VanBlaricom GR (2002) Food limitation and the recovery of sea otters following the '*Exxon Valdez*' oil spill. Mar Ecol Prog Ser 241:255–270

Driskell WB, Ruesink JL, Lees DC, Houghton JP, Lindstrom SC (2001) Long-term signal of disturbance: *Fucus gardneri* after the *Exxon Valdez* oil spill. Ecol Appl 11:815–827

Dzinbal KA (1982) Ecology of Harlequin Ducks in Prince William Sound, Alaska, during summer. M.S. thesis, Oregon State University, Corvallis

Dzinbal KA, Jarvis RL (1984) Coastal feeding ecology of Harlequin Ducks in Prince William Sound, Alaska, during summer. In: Nettleship DN, Sanger GA, Springer PF (eds) Marine birds: their feeding ecology and commercial fisheries relationships. Canadian Wildlife Service, Special Publication, Ottawa, pp 6–10

Ecological Consulting, Inc (1991) Assessment of direct seabird mortality in Prince William Sound and the western Gulf of Alaska resulting from the *Exxon Valdez* oil spill, Report (Bird Study Number 1) prepared for U.S. Department of Justice, Washington

Edgar B (1993) Has Alaska recovered? Am Birds 47:352–360

Esler D (2007) Abstract of study "Evaluating Harlequin Duck population recovery: CYP1A monitoring and a demographic population model," *Exxon Valdez* Oil Spill Trustee Council website. Exxon Valdez Oil Spill Trustee Council, Anchorage

Esler D (2008) Quantifying temporal variation in Harlequin Duck Cytochrome P4501A induction, *Exxon Valdez* oil spill Restoration Final Report (Restoration Project 050777). Exxon Valdez Oil Spill Trustee Council, Anchorage

Esler D, Iverson SA (2010) Female harlequin duck winter survival 11 to 14 years after the *Exxon Valdez* oil spill. J Wildl Mgmt 74:471–478

Esler D, Schmutz JA, Jarvis RL, Mulcahy DM (2000a) Winter survival of adult female Harlequin Ducks in relation to history of contamination by the *Exxon Valdez* oil spill. J Wildl Manage 64:839–847

Esler D, Bowman TD, Dean TA, O'Clair CE, Jewett SC, McDonald LL (2000b) Correlates of Harlequin Duck densities during winter in Prince William Sound, Alaska. Condor 102:920–926

Esler D, Bowman TD, Trust KA, Ballachey BE, Dean TA, Jewett SC, O'Clair CE (2002) Harlequin Duck population recovery following the '*Exxon Valdez*' oil spill: progress, process, and constraints. Mar Ecol Prog Ser 241:271–286

Esler DK, Trust KA, Iverson S, Mulcahy D (2003) Harlequin Ducks and the *Exxon Valdez* oil spill: the long road to recovery, Paper presented at SETAC 24th Annual Meeting in North America, Austin, TX, 9–13 Nov 2003, Abstract No. 573 in Abstract Book

Esler D, Trust KA, Ballachey BE, Iverson SA, Lewis TL, Rizzolo DJ, Mulcahy DM, Miles AK, Woodin BR, Stegeman JJ, Henderson JD, Wilson BW (2010) Cytochrome P4501A biomarker indication of oil exposure in harlequin ducks up to 20 years after the *Exxon Valdez* oil spill. Env Toxicol Chem DOI 10.1002/etc.129

Exxon Valdez Oil Spill Trustee Council (1994) *Exxon Valdez* oil spill restoration plan. Exxon Valdez Oil Spill Trustee Council, Anchorage

Exxon Valdez Oil Spill Trustee Council (1998) *Exxon Valdez* oil spill restoration plan. Exxon Valdez Oil Spill Trustee Council, Anchorage

Exxon Valdez Oil Spill Trustee Council (1999) *Exxon Valdez* oil spill restoration plan: update on injured resources and services. Exxon Valdez Oil Spill Trustee Council, Anchorage

Exxon Valdez Oil Spill Trustee Council (2001) 2001 status report. Exxon Valdez Oil Spill Trustee Council, Anchorage

Exxon Valdez Oil Spill Trustee Council (2002a) 2002 status report. Exxon Valdez Oil Spill Trustee Council, Anchorage

Exxon Valdez Oil Spill Trustee Council (2002b) *Exxon Valdez* oil spill restoration plan: update on injured resources and services, August, 2002. Exxon Valdez Oil Spill Trustee Council, Anchorage

Exxon Valdez Oil Spill Trustee Council (2006) *Exxon Valdez* oil spill restoration plan: update on injured resources and services, November, 2006. Exxon Valdez Oil Spill Trustee Council, Anchorage

Exxon Valdez Oil Spill Trustee Council (2009) 2009 Status Report. Exxon Valdez Oil Spill Trustee Council, Ancharge

Finney BP, Gregory-Eaves I, Douglas MSV, Smol JP (2002) Fisheries productivity in the northeastern Pacific Ocean over the past 2, 200 years. Nature 416:729–733

Fischer JB, Griffin CR (2000) Feeding behavior and food habits of wintering Harlequin Ducks at Shemya Island, Alaska. Wilson Bull 112:318–325

Forbes VE, Palmqvist A, Bach L (2006) The use and misuse of biomarkers in ecotoxicology. Environ Toxicol Chem 25:272–280

Francis RC, Hare SR, Hallowed AB, Wooster WS (1998) Effects of interdecadal climate variability on the oceanic ecosystems of the NE Pacific. Fish Oceanogr 7:1–21

Frink L, Ball-Weir K, Smith C (eds) (1995) Wildlife and oil spills. Tri-State Bird Rescue & Research, Newark

Furness RW, Greenwood JJD (eds) (1993) Birds as monitors of environmental change. Chapman & Hall, London

Gaines WL, Fitzner RE (1997) Winter diet of Harlequin Ducks at Sequim Bay, Puget Sound, Washington. Northwest Sci 61:213–215

Ganter B, Boyd H (2000) A tropical volcano, high predation pressure, and the breeding biology of arctic waterbirds: a circumpolar review of breeding failure in the summer of 1992. Arctic 53:289–305

Garrott RA, Eberhardt LL, Burn DM (1993) Mortality of sea otters in Prince William Sound following the *Exxon Valdez* oil spill. Mar Mamm Sci 9:343–359

Garshelis DL (1997) Sea otter mortality estimated from carcasses collected after the *Exxon Valdez* oil spill. Conserv Biol 11:905–916

Garshelis DL, Johnson CB (2001) Sea otter population dynamics and the *Exxon Valdez* oil spill: disentangling the confounding effects. J Anim Ecol 38:19–35

Gibeaut JC, Piper E (1998) 1993 shoreline oiling assessment of the *Exxon Valdez* oil spill, *Exxon Valdez* oil spill Final Report (Restoration Project 93038). Exxon Valdez Oil Spill Trustee Council, Anchorage

Gilfillan ES, Page DS, Harner EJ, Boehm PD (1995a) Shoreline ecology program for Prince William Sound, Alaska, following the *Exxon Valdez* oil spill: part 3 – biology. In: Wells PG, Butler JN, Hughes JS (eds) *Exxon Valdez* oil spill: fate and effects in Alaskan waters, ASTM Special Technical Publication No. 1219. American Society of Testing and Materials, Philadelphia, pp 398–443

Gilfillan ES, Suchanek TH, Boehm PD, Harner EJ, Page DS, Sloan NA (1995b) Shoreline impacts in the Gulf of Alaska region following the *Exxon Valdez* oil spill. In: Wells PG, Butler JN,

Hughes JS (eds) *Exxon Valdez* oil spill: fate and effects in Alaskan waters, ASTM Special Technical Publication No. 1219. American Society of Testing and Materials, Philadelphia, pp 444–481

Gilfillan ES, Harner EJ, Page DS (2002) Comment on Peterson et al. (2001) Sampling design begets conclusions. Mar Ecol Prog Ser 231:303–308

Golet GH, Seiser PE, McGuire AD, Roby DD, Fischer JB, Kuletz KJ, Irons DB, Dean TA, Jewett SC, Newman SH (2002) Long-term direct and indirect effects of the '*Exxon Valdez*' oil spill on pigeon guillemots in Prince William Sound, Alaska. Mar Ecol Prog Ser 241:287–304

Goudie RI (2006) Multivariate behavioural response of Harlequin Ducks to aircraft disturbance in Labrador. Environ Conserv 33:28–35

Goudie RI, Ankney CD (1986) Body size, activity budgets, and diets of sea ducks wintering in Newfoundland. Ecology 67:1475–1482

Goudie RI, Brault S, Conant B, Kondratyev AV, Petersen MR, Vermeer K (1994a) The status of sea ducks in the North Pacific Rim: toward their conservation and management. Trans N Am Wildl Nat Resour Conf 59:27–49

Goudie RI, Jones IL (2004) Dose-response relationships of Harlequin Duck behaviour to noise from low-level military jet over-flights in Central Labrador. Environ Conserv 31:289–298

Goudie RI, Lemon D, Brazil J (1994b) Observations of Harlequin Ducks, other waterfowl, and raptors in Labrador, 1987–1992, Canadian Wildlife Service Technical Report Series, No. 207. Atlantic Region, St. John's, NF, Canada.

Hahn ME (2002) Biomarkers and bioassays for detecting dioxin-like compounds in the marine environment. Sci Total Environ 289:49–69

Hare SR, Mantua NJ (2000) Empirical evidence for North Pacific regime shifts in 1977 and 1989. Prog Oceanogr 47:103–146

Harris PM, Rice SD, Babcock MM, Brodersen CC (1996) Within-bed distribution of *Exxon Valdez* crude oil in Prince William Sound blue mussels and underlying sediments. In: Rice SD, Spies RB, Wolfe DA, Wright BA (eds) Proceedings of the *Exxon Valdez* oil spill symposium, Symposium No. 18. American Fisheries Society, Bethesda, pp 298–308

Harwell MA, Gentile JH (2006) Ecological significance of residual exposures and effects from the *Exxon Valdez* oil spill. Integr Environ Assess Manage 2:204–246

Harwell MA, Gentile JH, Cummins KW, Highsmith RC, Hilborn R, McRoy CP, Parrish J, Weingartner T (2010) A conceptualization of the ecological risks to the Prince William Sound, Alaska, ecosystem. Hum Ecol Risk Assess (in press)

Harwell MA, Gentile JH, Johnson CB, Garshelis DL, Parker KR Quantitative ecological risk assessment of sea otters in Prince William Sound, Alaska: toxicological risks from *Exxon Valdez* subsurface oil residues. Hum Ecol Risk Assess (in press)

Hatch SA (2003) Statistical power for detecting trends with applications to seabird monitoring. Biol Conserv 111:317–329

Hayes MO, Michel J (1999) Factors determining the long-term persistence of *Exxon Valdez* oil in gravel beaches. Mar Pollut Bull 38:92–101

Hayward TL (1997) Pacific Ocean climate change: atmospheric forcing, ocean circulation, and ecosystem response. Trends Ecol Evol 12:150–154

Hoff RZ, Shigenaka G (1999) Lessons from 10 years old post-*Exxon Valdez* monitoring of inter-tidal shorelines, Proceedings of the 1999 International Oil Spill Conference. American Petroleum Institute, Seattle, pp 111–117

Holland-Bartels L, Ballachey B, Bishop MA, Bodkin J, Bowyer T, Dean T, Duffy L, Esler D, Jewett S, McDonald L, McGuire D, O'Clair C, Rebar A, Snyder P, Van Blaricom G (1998) Mechanisms of impact and potential recovery of nearshore vertebrate predators, *Exxon Valdez* oil spill Annual Report (Restoration Project 97025). Exxon Valdez Oil Spill Trustee Council, Anchorage

Holloway M (1996) Sounding out science. Sci Am 275(4):106–112

Hoover-Miller A, Parker KR, Burns JJ (2001) A reassessment of the impact of the *Exxon Valdez* oil spill on harbor seals (*Phoca vitulina richardsi*) in Prince William Sound, Alaska. Mar Mamm Sci 17:111–135

Hose JE, McGurk MD, Marty GD, Hinton DE, Brown ED, Baker TT (1996) Sublethal effects of the *Exxon Valdez* oil spill on herring embryos and larvae: morphological, cytogenetic, and histopathological assessments, 1989–1991. Can J Fish Aquat Sci 53:2355–2365

Huggett RJ, Stegeman JJ, Page DS, Parker KR, Woodin B, Brown JS (2003) Biomarkers in fish from Prince William Sound and the Gulf of Alaska: 1999–2000. Environ Sci Technol 37:4043–4051

Hunt GL Jr (1987) Offshore oil development and seabirds: the present status of knowledge and long-term research needs. In: Boesch DF, Rabalais NN (eds) Long-term environmental effects of offshore oil and gas development. Elsevier, New York, pp 539–586

Integral Consulting Inc (2006) Information Synthesis and Recovery Recommendations for Resources and Services Injured by the *Exxon Valdez* oil spill. Restoration Project 060783 Final Report. Integral Consulting, Mercer Island, WA 98040

Irons DB, Nysewander DR, Trapp JL (1988) Prince William Sound waterbird distribution in relation to habitat type. U.S. Fish and Wildlife Service, Anchorage

Irons DB, Kendall S, Erickson W, McDonald LL (1999) Chronic effects of the *Exxon Valdez* oil spill on summertime marine birds in Prince William Sound, Alaska. In: Abstracts, Legacy of an oil spill – 10 years after *Exxon Valdez*. Exxon Valdez Oil Spill Trustee Council, Anchorage, p 8

Irons DB, Kendall SJ, Erickson WP, McDonald LL, Lance BK (2000) Nine years after the *Exxon Valdez* oil spill: effects on marine bird populations in Prince William Sound, Alaska. Condor 102:723–737

Irons DB, Kendall SJ, Erickson WP, McDonald LL, Lance BK (2001) A brief response to Wiens et al., twelve years after the *Exxon Valdez* oil spill. Condor 103:892–894

Isleib ME, Kessel B (1973) Birds of the North Gulf Coast-Prince William Sound region, Alaska. Biol Pap Univ Alaska 14:1–149

Iverson SA, Esler D (2006) Site fidelity and the demographic implications of winter movements by a migratory bird, the Harlequin Duck *Histrionicus histrionicus*. J Avian Biol 37:219–228

Iverson SA, Esler D, Rizzolo DJ (2004) Winter philopatry of Harlequin Ducks in Prince William Sound, Alaska. Condor 106:711–715

Jewett SC, Dean TA, Woodin BR, Hoberg MK, Stegeman JJ (2002) Exposure to hydrocarbons 10 years after the *Exxon Valdez* oil spill: evidence from Cytochrome P4501A expression and biliary FACs in nearshore demersal fishes. Mar Environ Res 54:21–48

Jewett SC, Hoberg MK, Dean TA, Woodin BR, Stegeman JJ (2003) Authors' response. Mar Environ Res 55:463–488

Johnson CB, Garshelis DL (1995) Sea otter abundance, distribution, and pup production in Prince William Sound following the *Exxon Valdez* oil spill. In: (Wells PG, Butler JN, Hughes JS (eds) *Exxon Valdez* oil spill: fate and effects in Alaskan waters, ASTM Special Technical Publication No. 1219. American Society of Testing and Materials, Philadelphia, pp 894–929

Keeble J (1991) Out of the channel: the *Exxon Valdez* oil spill in Prince William Sound. HarperCollins, New York

Kenyon KW (1961) Birds of Amchitka Island, Alaska. Auk 78:305–326

King JG, Sanger GA (1979) Oil vulnerability index for marine oriented birds. In: Bartonek JC, Nettleship DN (eds) Conservation of marine birds of northern North America, Wildlife Research Report No. 11. U.S. Fish and Wildlife Service, pp 227–239

Kingston PF (2002) Long-term environmental impact of oil spills. Spill Sci Technol Bull 7:53–61

Klosiewski SP, Laing KK (1994) Marine bird populations of Prince William Sound, Alaska, before and after the *Exxon Valdez* oil spill, *Exxon Valdez* oil spill state/Federal Natural Resources Damage Assessment Final Report (Bird Study 2). Exxon Valdez Oil Spill Trustee Council, Anchorage

Knight GC, Walker CH (1982) A study of the hepatic microsomal monooxygenase of sea birds and its relationship to organochlorine pollutants. Comp Biochem Physiol 37C:211–221

Kocan RM, Hose JE, Brown ED, Baker TT (1996) Pacific herring (*Clupea pallasi*) sensitivity to Prudhoe Bay petroleum hydrocarbons: laboratory evaluation and in situ exposure at oiled and unoiled sites in Prince William Sound. Can J Fish Aquat Sci 53:2366–2375

Kuchel CR (1977) Some aspects of the behavior and ecology of Harlequin Ducks breeding in Glacier National Park, Montana. M.S. thesis, University of Montana, Missoula

Kvenvolden KA, Hostettler FD, Carlson PR, Rapp JB (1995) Ubiquitous tar balls with a California-source signature on the shorelines of Prince William Sound, Alaska. Environ Sci Technol 29:2684–2694

Lance BK, Irons DB, Kendall SJ, McDonald LL (1999) Marine bird and sea otter abundance of Prince William Sound, Alaska: trends following the T/V/*Exxon Valdez* oil spill, 1989–98, *Exxon Valdez* oil spill Restoration Project Annual Report (Restoration Project 98159). Exxon Valdez Oil Spill Trustee Council, Anchorage

Lance BK, Irons DB, Kendall SJ, McDonald LL (2001) An evaluation of marine bird population trends following the *Exxon Valdez* oil spill, Prince William Sound, Alaska. Mar Pollut Bull 42:298–309

Lanctot R, Goatcher B, Scribner K, Talbot S, Pierson B, Esler D, Zwiefelhofer D (1999) Harlequin Duck recovery from the *Exxon Valdez* oil spill: a population genetics perspective. Auk 116:781–791

Lee RF, Anderson JW (2005) Significance of Cytochrome P450 system responses and levels of bile fluorescent aromatic compounds in marine wildlife following oil spills. Mar Pollut Bull 50:705–723

Leighton FA (1993) The toxicity of petroleum to birds. Environ Rev 1:92–103

Longpré D, Jarry V, Fortin M-J (1997) Environmental impact sampling designs used to evaluate accidental spills. Spill Sci Technol Bull 4:133–139

Loughlin TR (ed) (1994) Marine mammals and the *Exxon Valdez*. Academic, San Diego

McCammon M (1999) Comments in the Conservation Roundtable, "Legacy of an oil spill." National Press Club of Washington, DC. 10 Feb 1999

McDonald TL, Erickson WP, McDonald LL (2000) Analysis of count data from before–after control–impact studies. J Agric Biol Environ Stat 5:262–279

McKnight A, Sullivan K, Irons DB, Stephenson SW, Howlin S (2006) Marine bird and sea otter population abundance of Prince William Sound, Alaska: trends following the T/V *Exxon Valdez* oil spill, 1989-2005, *Exxon Valdez* oil spill Restoration Project Annual Report (Restoration Projects 040159/050751). Exxon Valdez Oil Spill Trustee Council, Anchorage

Michel J, Nixon Z, Cotsapas L (2006) Evaluation of oil remediation technologies for lingering oil from the *Exxon Valdez* oil spill in Prince William Sound, Alaska. Restoration Project 050778 Final Report. Research Planning, Columbia, p 61

Mittelhauser GL (2000) The winter ecology of Harlequin Ducks in coastal Maine. M.S. Thesis, University of Maine, Orono

Monson DH, Doak DF, Ballachey BE, Johnson A, Bodkin JL (2000) Long-term impacts of the *Exxon Valdez* oil spill on sea otters, assessed through age-dependent mortality patterns. Proc Natl Acad Sci USA 97:6562–6567

Mudge SM (2002) Reassessment of the hydrocarbons in Prince William Sound and the Gulf of Alaska: identifying the source using partial least-squares. Environ Sci Technol 36:2354–2360

Murie OJ (1959) Fauna of the Aleutian Islands and Alaska Peninsula. N Am Fauna 61:1–406

Murphy SM, Day RH, Wiens JA, Parker KR (1997) Effects of the *Exxon Valdez* oil spill on birds: comparisons of pre- and post-spill surveys in Prince William Sound, Alaska. Condor 99:299–313

Murphy SM, Mabee TJ (2000) Status of black oystercatchers breeding in Prince William Sound, Alaska, 10 years after the *Exxon Valdez* oil spill. Waterbirds 23:204–213

National Wildlife Federation (2002) Prince William Sound: biological hot spots workshop report. National Wildlife Federation, Washington

National Wildlife Federation (2003) State of the sound: Prince William Sound, Alaska. National Wildlife Federation, Washington

Neff JM (1985) The use of biochemical measurements to detect pollutant-mediated damage to fish. In: Cardwell RD, Purdy R, Bahner RC (eds) Aquatic toxicology and hazard assessment: seventh symposium, ASTM STP854. American Society for Testing and Materials, Philadelphia, pp 155–183

Neff JM (2002) Bioaccumulation in marine organisms. Effects of contaminants from oil well produced water. Elsevier Science, Amsterdam

Neff JM, Owens EH, Stoker SW, McCormick DM (1995) Shoreline oiling conditions in Prince William Sound following the *Exxon Valdez* oil spill. In: Wells PG, Butler JN, Hughes JS (eds) *Exxon Valdez* oil spill: fate and effects in Alaskan waters, ASTM Special Technical Publication No. 1219. American Society of Testing and Materials, Philadelphia, pp 312–346

Neff JM, Stubblefield WA (1995) Chemical and toxicological evaluation of water quality following the *Exxon Valdez* oil spill. In: Wells PG, Butler JN, Hughes JS (eds) *Exxon Valdez* oil spill: fate and effects in Alaskan waters, ASTM Special Technical Publication No. 1219. American Society of Testing and Materials, Philadelphia, pp 141–177

Neff JM, Stout SA, Gunster DG (2005) Ecological risk assessment of PAHs in sediments, Identifying sources and toxicity. Integr Environ Assess Manage 1:22–33

Neff JM, Bence AE, Parker KR, Page DS, Brown JS, Boehm PD (2006) Bioavailability of polycyclic aromatic hydrocarbons from buried shoreline residues 13 years after the *Exxon Valdez* oil spill: a multispecies assessment. Environ Toxicol Chem 25:947–961

Nysewander D, Knudtson P (1977) The population ecology and migration of seabirds, shorebirds, and waterfowl associated with Constantine Harbor, Hinchinbrook Island, Prince William Sound, 1976. In: Bartonek JC, Lensink CJ, Gould RG [*sic*], Gill RE, Sanger GA (eds) Population dynamics and trophic relationships of marine birds in the Gulf of Alaska and southern Bering Sea. Environmental Assessment of the Alaskan Continental Shelf, Annual Reports, vol. 4. Outer Continental Shelf Environmental Assessment Program, Boulder, Colorado, pp 500–575

Oakley KL, Kuletz KJ (1979) Summer distribution and abundance of marine birds and mammals in the vicinity of Naked Island, Prince William Sound, Alaska, in 1978, and aspects of the reproductive ecology of the Pigeon Guillemot. U.S. Fish and Wildlife Service, Anchorage

O'Clair CE, Short JW, Rice SD (1996) Contamination of intertidal and subtidal sediments by oil from the *Exxon Valdez* in Prince William Sound. In: Rice SD, Spies RB, Wolfe DA, Wright BA (eds) Proceedings of the *Exxon Valdez* oil spill symposium, Symposium No. 18. American Fisheries Society, Bethesda, pp 61–93

Oris JT, Roberts AP (2007) Statistical analysis of cytochrome P4501A biomarker measurements in fish. Environ Toxicol Chem 26:1742–1750

Osenberg CW, Schmitt RJ, Holbrook SJ, Abu-Saba KE, Flegal AR (1994) Detection of environmental impacts: natural variability, effect size, and power analysis. Ecol Appl 4:16–30

Ott R (1999) *Exxon Valdez* aftermath. Defenders 74(2):16–24

Ott R (2005) Sound truth and corporate myth$: the legacy of the *Exxon Valdez* oil spill. Dragonfly Sisters, Cordova

Page DS, Boehm PD, Douglas GS, Bence AE, Burns WA, Mankiewicz PJ (1999) Pyrogenic polycyclic aromatic hydrocarbons in sediments record past human activity: a case study in Prince William Sound, Alaska. Mar Pollut Bull 38:247–260

Page DS, Gilfillan ES, Boehm PD, Harner EJ (1995) Shoreline Ecology Program for Prince William Sound, Alaska, following the *Exxon Valdez* oil spill: part 1 – study design and methods. In: Wells PG, Butler JN, Hughes JS (eds) *Exxon Valdez* oil spill: fate and effects in Alaskan waters, ASTM Special Technical Publication No. 1219. American Society of Testing and Materials, Philadelphia, pp 263–295

Page DS, Boehm PD, Stubblefield WA, Parker KR, Gilfillan ES, Neff JM, Maki AW (2002a) Hydrocarbon composition and toxicity of sediments following the *Exxon Valdez* oil spill in Prince William Sound, Alaska, USA. Environ Toxicol Chem 21:1438–1450

Page DS, Gilfillan ES, Boehm PD, Neff JM, Stubblefield WA, Parker KR, Maki AW (2002b) Sediment toxicity measurements in oil spill injury assessment: a study of shorelines affected by the *Exxon Valdez* oil spill in Prince William Sound, Alaska. In: Pellei M, Porta A, Hinchee RE (eds) Proceedings of the First International Conference on Remediation of Contaminated Sediments, Venice, Italy, 10–11 Oct 2001. Battelle, Columbus, pp 137–146

Page DS, Gilfillan ES, Boehm PD, Stubblefield WA, Parker KR, Neff JM, Maki AW (2003) Authors' reply. Environ Toxicol Chem 22:2539–2540

Page DS, Boehm PD, Brown JS, Neff JM, Burns WA, Bence AE (2005) Mussels document loss of bioavailable polycyclic aromatic hydrocarbons and the return to baseline conditions for oiled shorelines in Prince William Sound, Alaska. Mar Environ Res 60:422–436

Page DS, Brown JS, Boehm PD, Bence AE, Neff JM (2006) A hierarchical approach measures the aerial extent and concentration levels of PAH-contaminated shoreline sediments at historic industrial sites in Prince William Sound, Alaska. Mar Pollut Bull 52:367–379

Page DS, Boehm PD, Neff JM (2008) Shoreline type and subsurface oil persistence in the *Exxon Valdez* spill zone of Prince William Sound, Alaska. In: Proceedings of the 31st AMOP Technical Seminar on Environmental Contamination and Response, Environment Canada, Calgary, pp 545–564

Pain S (1993) Species after species suffers from Alaska's spill. New Sci 137(1860), 13 Feb: 5–6

Paine RT, Ruseink JL, Sun A, Soulanille EL, Wonham MJ, Harley CDG, Brumbaugh DR, Secord DL (1996) Trouble on oiled waters: lessons from the *Exxon Valdez* oil spill. Annu Rev Ecol Syst 27:197–235

Parker KR, Wiens JA (2005) Assessing recovery following environmental accidents: environmental variation, ecological assumptions, and strategies. Ecol Appl 15:2037–2051

Patten SM Jr (1993a) Acute and sublethal effects of the *Exxon Valdez* oil spill on Harlequins and other seaducks, in: *Exxon Valdez* oil spill symposium, 2–5 Feb 1993, book of extended abstracts. Exxon Valdez Oil Spill Trustee Council, Anchorage, pp 151–154

Patten SM Jr (1993b) Reproductive failure of Harlequin Ducks. Alaska Wildl 25:14–15

Patten SM Jr (1994) Assessment of injury to sea ducks from hydrocarbon uptake in Prince William Sound and the Kodiak Archipelago, Alaska, following the *Exxon Valdez* oil spill, *Exxon Valdez* oil spill state/Federal Natural Resource Damage Assessment Final Report (NRDA Bird Study 11). Exxon Valdez Oil Spill Trustee Council, Anchorage

Patten SM Jr, Crowley DW, Crowe TW, Gustin R (1995) Restoration monitoring of Harlequin Ducks (*Histrionicus histrionicus*) in Prince William Sound and Afognak Island, *Exxon Valdez* oil spill Restoration Project Draft Report (Restoration Project 93-033). Exxon Valdez Oil Spill Trustee Council, Anchorage

Patten SM Jr, Crowe T, Gustin R, Hunter R, Twait P, Hastings C (2000) Assessment of injury to sea ducks from hydrocarbon uptake in Prince William Sound and the Kodiak Archipelago, Alaska, following the *Exxon Valdez* oil spill, *Exxon Valdez* oil spill state/Federal Natural Resource Damage Assessment Final Report (NRDA Bird Study 11), Vols I and II. Exxon Valdez Oil Spill Trustee Council, Anchorage

Pearson WH, Elston RA, Bienert RW, Drum AS, Antrim LD (1999) Why did the Prince William Sound, Alaska, Pacific herring (*Clupea pallasi*) fisheries collapse in 1993 and 1994? Review of hypotheses. Can J Fish Aquat Sci 56:711–737

Perfito N, Schirato G, Brown M, Wingfield JC (2002) Response to acute stress in the Harlequin Duck (*Histrionicus histrionicus*) during the breeding season and moult: relationships to gender, condition, and life-history stage. Can J Zool 80:1334–1343

Peterson CH (2001) The "*Exxon Valdez*" oil spill in Alaska: acute, indirect and chronic effects on the ecosystem. Adv Mar Biol 39:1–103

Peterson CH, McDonald LL, Green RH, Erickson WP (2001) Sampling design begets conclusions: the statistical basis for detection of injury to and recovery of shoreline communities after the '*Exxon Valdez*' oil spill. Mar Ecol Prog Ser 210:255–283

Peterson CH, McDonald LL, Green RH, Erickson WP (2002) The joint consequences of multiple components of statistical sampling design. Mar Ecol Prog Ser 231:309–314

Peterson CH, Rice SD, Short JW, Esler D, Bodkin JL, Ballachey BE, Irons DB (2003) Long-term ecosystem response to the *Exxon Valdez* oil spill. Science 302:2082–2086

Peterson WT, Schwing FB (2003) A new climate regime in northeast pacific [sic] ecosystems. Geophys Res Lett 30(17):4 pp. Available at: doi:10.1029/2003GL017528

Piatt JF, Anderson P (1996) Responses of Common Murres to the *Exxon Valdez* oil spill and long-term changes in the Gulf of Alaska marine ecosystem. In: Rice SD, Spies RB, Wolfe DA, Wright BA (eds) Proceedings of the *Exxon Valdez* oil spill symposium, Symposium No. 18. American Fisheries Society, Bethesda, pp 720–737

Piatt JF, Lensink CJ, Butler W, Kendziorek M, Nysewander DR (1990) Immediate impact of the *Exxon Valdez* oil spill on marine birds. Auk 107:387–397

Piatt JF, Ford RG (1996) How many seabirds were killed by the *Exxon Valdez* oil spill? In: Rice SD, Spies RB, Wolfe DA, Wright BA (eds) Proceedings of the *Exxon Valdez* oil spill symposium, Symposium No. 18. American Fisheries Society, Bethesda, pp 712–719

Platt JR (1964) Strong inference. Science 146:347–353

Rattner BA, Hoffman DJ, Marn CM (1989) Use of mixed-function oxygenases to monitor contaminant exposure in wildlife. Environ Toxicol Chem 8:1093–1102

Rice SD, Spies RB, Wolfe DA, Wright BA (eds) (1996) Proceedings of the *Exxon Valdez* oil spill symposium. American Fisheries Society, Bethesda, Symposium No. 18

Rice SD, Thomas RE, Carls MG, Heintz RA, Wertheimer AC, Murphy ML, Short JW, Moles A (2001) Impacts to pink salmon following the *Exxon Valdez* oil spill: persistence, toxicity, sensitivity, and controversy. Rev Fish Sci 9:165–211

Rice SD, Carls MG, Heintz RA, Short JW (2003) Comment on "Hydrocarbon composition and toxicity of sediments following the *Exxon Valdez* oil spill in Prince William Sound, Alaska, USA". Environ Toxicol Chem 11:2539–2540

Robertson GJ (1997) Pair formation, mating system, and winter philopatry in Harlequin Ducks, Ph.D. dissertation. Simon Fraser University, Burnaby

Robertson GJ, Goudie RI (1999) Harlequin Duck (*Histrionicus histrionicus*). In: Gill F, Poole A (eds) The birds of North America, No. 466. Birds of North America, Philadelphia

Robertson GJ, Cooke F, Goudie RI, Boyd WS (2000) Spacing patterns, mating systems, and winter philopatry in Harlequin Ducks. Auk 117:299–307

Rodway MS, Regehr HM, Ashley J, Clarkson PV, Goudie RI, Hay DE, Smith CM, Wright KG (2003a) Aggregative response of Harlequin Ducks to herring spawning in the Strait of Georgia, British Columbia. Can J Zool 81:504–514

Rodway MS, Regehr HM, Cooke F (2003b) Sex and age differences in distribution, abundance, and habitat preferences of wintering Harlequin Ducks: implications for conservation and estimating recruitment rates. Can J Zool 81:492–503

Rosenberg D (2008) Harlequin duck population dynamics in Prince William Sound: Measuring recovery. 2008 Progress Report, Project 080759. Exxon Valdez Oil Spill Trustee Council, Anchorage

Rosenberg DH, Petrula MJ (1998) Status of Harlequin Ducks in Prince William Sound, Alaska after the *Exxon Valdez* oil spill, 1995–1997, *Exxon Valdez* oil spill Restoration Project Final Report (Restoration Project 97427). Exxon Valdez Oil Spill Trustee Council, Anchorage

Rosenberg DH, Petrula MJ (1999) Recovery monitoring of Harlequin Ducks in Prince William Sound, Alaska. In: Abstracts. Legacy of an oil spill – 10 years after *Exxon Valdez*. *Exxon Valdez* Oil Spill Trustee Council, Anchorage, p. 5

Rosenberg DH, Petrula MJ (2001) Status of Harlequin Ducks in Prince William Sound, Alaska after the *Exxon Valdez* oil spill, 1995–1997, *Exxon Valdez* oil spill Restoration Project Final Report (Restoration Project 97427). Exxon Valdez Oil Spill Trustee Council, Anchorage

Rosenberg D, Petrula MP (2005) Recruitment rates in Harlequin Ducks: when do we become concerned? Abstracts of the Second North American Sea Duck Conference, Annapolis, p 21

Rosenberg DH, Petrula MJ, Hill DD, Christ AM (2005) Harlequin Duck population dynamics: measuring recovery from the *Exxon Valdez* oil spill, *Exxon Valdez* oil spill Restoration Project Final Report (Restoration Project 04047 [through 05047]). Exxon Valdez Oil Spill Trustee Council, Anchorage

Royer TC, Vermersch JA, Weingartner TJ, Niebauer HJ, Muench RD (1990) Ocean circulation influencing the *Exxon Valdez* oil spill. Oceanography 3:3–10

Sangster ME, Kurhajec DJ, Benz CT (1978) Reproductive ecology of seabirds at Hinchinbrook Island and a census of seabirds at selected sites in Prince William Sound, 1977. U.S. Fish and Wildlife Service, Anchorage

Schmitt RJ, Osenberg CW (1996) Detecting ecological impacts: concepts and applications in coastal habitats. Academic, San Diego

Shigenaka G, Henry CB Jr (1993) Bioavailability of residual PAHs from the *Exxon Valdez* oil spill. In: *Exxon Valdez* oil spill symposium, 2–5 Feb 1993, book of extended abstracts. Anchorage, pp 163–165

Short JW, Harris PM (1996) Petroleum hydrocarbons in caged mussels deployed in Prince William Sound after the *Exxon Valdez* oil spill. In: Rice SD, Spies RB, Wolfe DA, Wright BA (eds) Proceedings of the *Exxon Valdez* oil spill symposium, Symposium No. 18. American Fisheries Society, Bethesda, pp 29–39

Short JW, Lindeberg MR, Harris PM, Maselko J, Rice SD (2002) Vertical oil distribution within the intertidal zone 12 years after the *Exxon Valdez* oil spill in Prince William Sound, Alaska. In: Proceedings of the 25th Arctic and Marine Oil Spill Program (AMOP) Technical Seminar. Environment Canada, Ottawa, pp 57–72

Short JW, Lindeberg MR, Harris PM, Maselko JM, Pella JJ, Rice SD (2004) Estimate of oil persisting on the beaches of Prince William Sound 12 years after the *Exxon Valdez* oil spill. Environ Sci Technol 38:19–25

Short JW, Maseklo JM, Lindeberg MR, Harris PM, Rice SD (2006) Vertical distribution and probability of encountering intertidal *Exxon Valdez* oil on shorelines of three embayments within Prince William Sound. Environ Sci Technol 40:3723–3729

Short JW, Irvine GV, Mann DH, Maselko JM, Pella JJ, Lindeberg MR, Payne JR, Driskell WB, Rice SD (2007) Slightly weathered Exxon Valdex oil persists in Gulf of Alaska beach sediments after 16 years. Environ Sci Technol 41:1245–1250

Shrader-Frechette KS, McCoy ED (1993) Methods in ecology: strategies for conservation. Cambridge University, Cambridge

Skalski JR (1995) Statistical considerations in the design and analysis of environmental damage assessment studies. J Environ Manage 43:67–85

Skalski JR, Coats DA, Fukuyama AK (2001) Criteria for oil spill recovery: a case study of the intertidal community of Prince William Sound, Alaska, following the *Exxon Valdez* oil spill. Environ Manage 28:9–18

Stegeman JJ, Hahn ME (1994) Biochemistry and molecular biology of monooxygenases: current perspectives on forms, functions, and regulation of cytochrome P450 in aquatic species. In: Malins DC, Ostrander GK (eds) Aquatic toxicology: molecular, biochemical and cellular perspectives. CRC/Lewis, Boca Raton, pp 87–206

Stephensen SW, Irons DB, Kendall SJ, Lance BK, McDonald LL (2001) Marine bird and sea otter population abundance of Prince William Sound, Alaska: trends following the *T/V Exxon Valdez* oil spill, 1989–2000, *Exxon Valdez* oil spill Restoration Project Annual Report (Restoration Project 00159). Exxon Valdez Oil Spill Trustee Council, Anchorage

Stubblefield WA, Hancock GA, Ford WH, Prince HH, Ringer RK (1995a) Evaluation of the toxic properties of naturally weathered *Exxon Valdez* crude oil to surrogate wildlife species. In: Wells PG, Butler JN, Hughes JS (eds) *Exxon Valdez* oil spill: fate and effects in Alaskan waters, ASTM Special Technical Publication No. 1219. American Society of Testing and Materials, Philadelphia, pp 665–692

Stubblefield WA, Hancock GA, Prince HH, Ringer RK (1995b) Effects of naturally weathered *Exxon Valdez* crude oil on Mallard reproduction. Environ Toxicol Chem 14:1951–1960

Taylor E, Reimer D (2008) Oil persistence on beaches in Prince William Sound – a review of SCAT surveys conducted from 1989 to 2002. Mar Pollut Bull 56:458–474

Trust KA, Esler D, Woodin BR, Stegeman JJ (2000) Cytochrome P450 1A induction in sea ducks inhabiting nearshore areas of Prince William Sound, Alaska. Mar Pollut Bull 40:397–403

Ver Hoef JM, Frost KJ (2003) A Bayesian hierarchical model for monitoring harbor seal changes in Prince William Sound, Alaska. Environ Ecol Stat 10:201–219

Vermeer K (1983) Diet of the Harlequin Duck in the Strait of Georgia, British Columbia. Murrelet 64:54–57

Wallen RL (1987) Habitat utilization by Harlequin Ducks in Grant Teton National Park. M.S. thesis, Montana State University, Bozeman

Wells PG, Butler JN, Hughes JS (eds) (1995) *Exxon Valdez* oil spill: fate and effects in Alaskan waters. American Society for Testing and Materials, Philadelphia, PA, ASTM Special Technical Publication No. 1219

Wheelwright J (1994) Degrees of disaster – Prince William Sound: how Nature reels and rebounds. Simon and Schuster, New York

Widdows J, Donkin PD, Evans SV, Page DS, Salkeld PN (1995) Sublethal biological effects and chemical contaminant monitoring of Sullom Voe (Shetland) using mussels (*Mytilus edulis*). Proc R Soc Edinb B 103:99–112

Wiens JA (1977) On competition and variable environments. Am Sci 65:590–597

Wiens JA (1995) Recovery of seabirds following the *Exxon Valdez* oil spill: an overview. In: Wells PG, Butler JN, Hughes JS (eds) *Exxon Valdez* oil spill: fate and effects in Alaskan waters, ASTM Special Technical Publication No. 1219. American Society of Testing and Materials, Philadelphia, pp 854–893

Wiens JA (1996) Oil, seabirds, and science: the effects of the *Exxon Valdez* oil spill. Bioscience 46:587–597

Wiens JA (1997) Scientific responsibility and responsible ecology. Conserv Ecol 1(1):16. Available online: http://www.consecol.org/vol1/iss1/art16

Wiens JA (2007) Applying ecological risk assessment to environmental accidents: Harlequin ducks and the *Exxon Valdez* oil spill. Bioscience 57:769–777

Wiens JA (2008) Uncertainty and the relevance of ecology. Bull Br Ecol Soc 39(2):47–48

Wiens JA, Crist TO, Day RH, Murphy SM, Hayward GD (2001a) A canonical correspondence analysis of the effects of the *Exxon Valdez* oil spill on marine birds. Ecol Appl 11:828–839

Wiens JA, Day RH, Murphy SM, Parker KR (2001b) On drawing conclusions nine years after the *Exxon Valdez* oil spill. Condor 103:886–892

Wiens JA, Day RH, Murphy SM, Parker KR (2004) Changing habitat and habitat use by birds after the *Exxon Valdez* oil spill, 1989–2001. Ecol Appl 14:806–1825

Wiens JA, Parker KR (1995) Analyzing the effects of accidental environmental impacts: approaches and assumptions. Ecol Appl 5:1069–1083

Wolfe DA, Hameedi MJ, Galt JA, Watabayashi G, Short J, O'Clair C, Rice S, Michel J, Payne JR, Braddock J, Hanna S, Sale D (1994) Fate of the oil spilled from the *Exxon Valdez*. Environ Sci Technol 28:561–568

Wolfe DA, Krahn MM, Casillas E, Sol S, Thompson TA, Lunz J, Scott KJ (1996) Toxicity of intertidal and subtidal sediments contaminated by the *Exxon Valdez* oil spill. In: Rice SD, Spies RB, Wolfe DA, Wright BA (eds) Proceedings of the *Exxon Valdez* oil spill symposium, Symposium No. 18. American Fisheries Society, Bethesda, pp 121–139

Wooley C (2002) The myth of the "pristine environment": past human impacts in Prince William Sound and the Northern Gulf of Alaska. Spill Sci Technol Bull 7:89–104

Index

Printed by Printforce, the Netherlands